Hugo Hens

Performance Based Building Design 1

From Below Grade Construction to Cavity Walls

Hugo Hens

Performance Based Building Design 1

From Below Grade Construction to Cavity Walls

Professor Hugo S .L. C. Hens
University of Leuven (KULeuven)
Department of Civil Engineering
Building Physics
Kasteelpark Arenberg 40
3001 Leuven
Belgium

Coverphoto: © Hugo Hens

Library of Congress Card No.:
applied for

British Library Cataloguing-in-Publication Data
A catalogue record for this book is available from the British Library.

Bibliographic information published by the Deutsche Nationalbibliothek
The Deutsche Nationalbibliothek lists this publication in the Deutsche Nationalbibliografie; detailed bibliographic data are available on the Internet at http://dnb.d-nb.de.

© 2012 Wilhelm Ernst & Sohn
Verlag für Architektur und technische Wissenschaften GmbH & Co. KG, Rotherstr. 21, 10245 Berlin, Germany

All rights reserved (including those of translation into other languages). No part of this book may be reproduced in any form – by photoprinting, microfilm, or any other means – nor transmitted or translated into a machine language without written permission from the publishers. Registered names, trademarks, etc. used in this book, even when not specifically marked as such, are not to be considered unprotected by law.

Coverdesign: Sophie Bleifuß, Berlin, Germany
Typesetting: Manuela Treindl, Fürth, Germany
Printing and Binding: betz-druck GmbH, Darmstadt, Germany

Printed in the Federal Republic of Germany.
Printed on acid-free paper.

Print ISBN: 978-3-433-03022-6
ePDF ISBN: 978-3-433-60196-9
ePub ISBN: 978-3-433-60197-6
mobi ISBN: 978-3-433-60198-3
oBook ISBN: 978-3-433-60195-2

To my wife, children and grandchildren

In remembrance of Professor A. de Grave
Who introduced building physics as a new discipline
at the University of Leuven (KULeuven), Belgium, in 1952

Contents

	Preface	XIII
0	**Introduction**	1
0.1	Subject of the book	1
0.2	Units and symbols	1
0.3	References and literature	5
1	**Performances**	7
1.1	In general	7
1.2	Definitions and basic characteristics	7
1.3	Advantages	7
1.4	Performance arrays	7
1.5	Design based on performance metrics	10
1.5.1	The design process	10
1.5.2	Integrating a performance analysis	10
1.6	Impact on the building process	11
1.7	References and literature	11
2	**Materials**	13
2.1	In general	13
2.2	Array of material properties	13
2.3	Thermal insulation materials	14
2.3.1	Introduction	14
2.3.2	Apparent thermal conductivity	14
2.3.2.1	In general	14
2.3.2.2	Impact of the transport modes	14
2.3.3	Other properties	19
2.3.3.1	Mechanical	19
2.3.3.2	Physical	19
2.3.3.3	Fire	20
2.3.3.4	Sensitivity to temperature, IR and UV	20
2.3.4	Materials	20
2.3.4.1	Insulating building materials	20
2.3.4.2	Insulation materials	23
2.3.4.3	Insulating systems	34
2.3.4.4	Recent developments	35
2.4	Water, vapour and air flow control layers	37
2.4.1	In general	37

2.4.2	Water barriers	38
2.4.2.1	A short history	38
2.4.2.2	Bituminous membranes	39
2.4.2.3	Polymer-bituminous membranes	39
2.4.2.4	High-polymer membranes	41
2.4.3	Vapour retarders and vapour barriers	42
2.4.4	Air barriers	44
2.5	Joints	44
2.5.1	In general	44
2.5.2	Joint solutions and joint finishing options	45
2.5.3	Performance requirements	46
2.5.3.1	Mechanical	46
2.5.3.2	Building physics related	46
2.5.4	Sealant classification	46
2.5.5	Load and sealant choice	48
2.5.6	Structural design of sealed joints	49
2.5.7	Points of attention	50
2.6	References and literature	51
3	**Excavations and building pit**	**55**
3.1	In general	55
3.2	Realisation	55
4	**Foundations**	**57**
4.1	In general	57
4.2	Performance evaluation	57
4.2.1	Structural integrity	57
4.2.1.1	Equilibrium load bearing capacity	57
4.2.1.2	Settling load bearing capacity	58
4.2.2	Building physics	60
4.2.3	Durability	60
4.3	Foundation systems	61
4.3.1	In general	61
4.3.2	Spread foundations	61
4.3.2.1	Footings	61
4.3.2.2	Foundation slabs	63
4.3.2.3	Soil consolidation	63
4.3.3	Deep foundations	63
4.3.3.1	Wells	63
4.3.3.2	Piles	64
4.4	Specific problems	65
4.4.1	Eccentrically loaded footings	65
4.4.2	Footings under large openings	66
4.4.3	Reinforcing and/or deepening existing foundations	66
4.4.3.1	Footings	66

4.4.3.2	Wells	67
4.4.3.3	Pressed piles	67
4.5	References and literature	68
5	**Building parts on and below grade**	**69**
5.1	In general	69
5.2	Performance evaluation	69
5.2.1	Structural integrity	69
5.2.1.1	Static stability	69
5.2.1.2	Strength and stiffness	70
5.2.2	Building physics, heat, air, moisture	71
5.2.2.1	Air tightness	71
5.2.2.2	Thermal transmittance	73
5.2.2.3	Transient response	88
5.2.2.4	Moisture tolerance	91
5.2.2.5	Thermal bridging	98
5.2.3	Building physics: acoustics	100
5.2.4	Durability	101
5.2.5	Fire safety	101
5.2.6	Soil gases	101
5.3	Design and execution	101
5.3.1	Basements	101
5.3.2	Drainages	102
5.3.2.1	In general	102
5.3.2.2	Properties	103
5.3.2.3	Design	103
5.3.3	Waterproof encasement	105
5.3.3.1	Inside	105
5.3.3.2	Outside	107
5.3.4	Waterproof concrete	108
5.4	References and literature	109
6	**Structural options**	**111**
6.1	In general	111
6.2	Performance evaluation	112
6.2.1	Structural integrity	112
6.2.2	Fire safety	113
6.3	Structural system design	115
6.3.1	Vertical loads	115
6.3.2	Horizontal load	116
6.3.2.1	Massive structures	116
6.3.2.2	Skeleton structures	119
6.3.3	Dynamic horizontal loads	121
6.4	References and literature	121

7	**Floors**	123
7.1	In general	123
7.2	Performance evaluation	124
7.2.1	Structural integrity	124
7.2.2	Building physics: heat-air-moisture	125
7.2.2.1	Air tightness	125
7.2.2.2	Thermal transmittance	126
7.2.2.3	Transient response	128
7.2.2.4	Moisture tolerance	129
7.2.2.5	Thermal bridging	135
7.2.3	Building physics: acoustics	136
7.2.3.1	Airborne noise	136
7.2.3.2	Impact noise	136
7.2.4	Durability	137
7.2.5	Fire safety	138
7.3	Design and execution	139
7.3.1	In general	139
7.3.2	Timber floors	140
7.3.2.1	Span below 6 m	140
7.3.2.2	Spans above 6 m	142
7.3.3	Concrete slabs and prefabricated structural floor units	142
7.3.3.1	Span below 6 m	142
7.3.3.2	Span above 6 m	145
7.3.4	Steel floors	146
7.3.4.1	Span below 6 m	146
7.3.4.2	Span above 6 m	146
7.4	References and literature	147
8	**Outer wall requirements**	149
8.1	In general	149
8.2	Performance evaluation	149
8.2.1	Structural integrity	149
8.2.2	Building physics: heat, air, moisture	150
8.2.2.1	Air tightness	150
8.2.2.2	Thermal transmittance	151
8.2.2.3	Transient response	152
8.2.2.4	Moisture tolerance	153
8.2.2.5	Thermal bridging	153
8.2.3	Building physics: acoustics	153
8.2.4	Durability	154
8.2.5	Fire safety	155
8.2.6	Maintenance and economy	155
8.3	References and literature	155

9 Massive outer walls ... 157

9.1 Traditional masonry walls ... 157
9.1.1 In general ... 157
9.1.2 Performance evaluation ... 157
9.1.2.1 Building physics: heat, air, moisture ... 157
9.1.2.2 Building physics: acoustics ... 160
9.1.2.3 Durability ... 160
9.1.2.4 Fire safety ... 160
9.1.3 Conclusion ... 160

9.2 Massive light-weight walls ... 160
9.2.1 In general ... 160
9.2.2 Performance evaluation ... 161
9.2.2.1 Structural integrity ... 161
9.2.2.2 Building physics: heat, air, moisture ... 162
9.2.2.3 Building physics: acoustics ... 170
9.2.2.4 Durability ... 171
9.2.2.5 Fire safety ... 172
9.2.2.6 Maintenance ... 172
9.2.3 Design and execution ... 172
9.2.3.1 In general ... 172
9.2.3.2 Specific ... 173

9.3 Massive walls with inside insulation ... 174
9.3.1 In general ... 174
9.3.2 Performance evaluation ... 174
9.3.2.1 Structural integrity ... 174
9.3.2.2 Building physics: heat, air, moisture ... 174
9.3.2.3 Building physics: acoustics ... 190
9.3.2.4 Durability ... 190
9.3.2.5 Fire safety ... 191
9.3.2.6 Global conclusion ... 192
9.3.3 Design and execution ... 192

9.4 Massive walls with outside insulation ... 194
9.4.1 In general ... 194
9.4.2 Performance evaluation ... 195
9.4.2.1 Structural integrity ... 195
9.4.2.2 Building physics: heat, air, moisture ... 195
9.4.2.3 Building physics: acoustics ... 206
9.4.2.4 Durability ... 207
9.4.2.5 Fire safety ... 209
9.4.2.6 Maintenance ... 209
9.4.2.7 Global conclusion ... 209
9.4.3 Design and execution ... 209
9.4.3.1 Clad stud systems ... 209
9.4.3.2 EIFS-systems ... 210

9.5 References and literature ... 212

10	**Cavity walls**	215
10.1	In general	215
10.2	Performance evaluation	217
10.2.1	Structural integrity	217
10.2.2	Building physics: heat, air, moisture	218
10.2.2.1	Air tightness	218
10.2.2.2	Thermal transmittance	221
10.2.2.3	Transient response	234
10.2.2.4	Moisture tolerance	235
10.2.2.5	Thermal bridges	244
10.2.3	Building physics: acoustics	244
10.2.4	Durability	245
10.2.1	Fire safety	246
10.2.1	Maintenance	246
10.3	Design and execution	247
10.3.1	New construction	247
10.3.1.1	Airtight, as few thermal bridges as possible	247
10.3.1.2	Correct cavity trays where needed	248
10.3.1.3	Excluding air looping and wind washing	249
10.3.2	Post-filling existing cavity walls	250
10.4	References and literature	251
11	**Panelized massive outer walls**	255
11.1	In general	255
11.2	Performance evaluation	256
11.2.1	Structural integrity	256
11.2.2	Building physics: heat, air, moisture	257
11.2.2.1	Air tightness	257
11.2.2.2	Thermal transmittance	257
11.2.2.3	Transient response	259
11.2.2.4	Moisture tolerance	259
11.2.2.5	Thermal bridging	260
11.2.3	Building physics: acoustics	260
11.2.4	Durability	260
11.2.5	Fire safety	261
11.2.6	Maintenance	261
11.3	Design and execution	261
11.4	References and literature	262

Preface

Overview

Just like building physics, performance based building design was hardly an issue before the energy crises of the nineteen seventies. Together with the need for more energy efficiency, the interest in overall building performance grew. The tome on applied building physics already discussed a performance rationale, and contained an in depth analysis of the heat, air, moisture performance requirements at the building and building enclosure level. This third tome builds on that rationale although also structural aspects, acoustics, fire safety, maintenance and buildability are considered now. The text reflects thirty eight years of teaching architectural, building and civil engineers, coupled to more than forty years of experience in research and consultancy. Where and when needed, input from over the world was used, reason why each chapter ends with a list of references and literature.

The book should be usable by undergraduates and graduates in architectural and building engineering, though also building engineers, who want to refresh their knowledge, may benefit. The level of discussion assumes the reader has a sound knowledge of building physics, along with a background in structural engineering, building materials and building construction.

Acknowledgments

A book of this magnitude reflects the work of many, not only of the author. Therefore, first of all, we like to thank the thousands of students we had. They gave us the opportunity to test the content and helped in upgrading it.

The text should not been written the way it is, if not standing on the shoulders of those, who preceded. Although we started our carrier as a structural engineer, our predecessor, Professor Antoine de Grave, planted the seeds that slowly fed our interest in building physics, building services and performance based building design. The late Bob Vos of TNO, the Netherlands, and Helmut Künzel of the Fraunhofer Institut für Bauphysik, Germany, showed the importance of experimental work and field testing for understanding building performance, while Lars Erik Nevander of Lund University, Sweden, taught that application does not always ask extended modeling, mainly because reality in building construction is much more complex than any simulation can reflect.

During the four decennia at the Laboratory of Building Physics, several researchers and PhD-students got involved. I am very grateful to Gerrit Vermeir, Staf Roels Dirk Saelens and Hans Janssen who became colleagues at the university; to Jan Carmeliet, now professor at the ETH-Zürich; Piet Standaert, a principal at Physibel Engineering; Jan Lecompte, at Bekaert NV; Filip Descamps, a principal at Daidalos Engineering and part-time professor at the Free University Brussels (VUB); Arnold Janssens, associate professor at the University of Ghent (UG); Rongjin Zheng, associate professor at Zhejiang University, China, Bert Blocken, professor at the Technical University Eindhoven (TU/e) and Griet Verbeeck, professor at KHL, who all contributed by their work. The experiences gained by working as a structural engineer and building site supervisor at the start of my career, as building assessor over the years, as researcher and operating agent of four Annexes of the IEA, Executive Committee on Energy Conservation in Buildings and Community Systems forced me to rethink the engineering based performance approach every time again. The many ideas I exchanged and got in Canada and the USA from Kumar Kumaran, Paul Fazio, Bill Brown, William B. Rose, Joe Lstiburek and

Anton Ten Wolde were also of great help. A number of reviewers took time to examine the book. Although we do not know their names, we like to thank them.

Finally, I thank my family, my wife Lieve, who managed living together with a busy engineering professor, my three children who had to live with that busy father and my many grandchildren who do not know their grandfather is still busy.

Leuven, February 2012 *Hugo S. L. C. Hens*

0 Introduction

0.1 Subject of the book

This is the third book in a series on building physics, applied building physics and performance based building design:

- Building Physics: Heat, Air and Moisture
- Applied Building Physics: Boundary Conditions, Building Performance and Material Properties
- **Performance Based Building Design 1**
- Performance Based Building Design 2

Both volumes apply the performance based engineering rationale, discussed in 'Applied Building Physics: Boundary Conditions, Building Performance and Material Properties' to the design and construction of building elements and assemblies. In order to do that, the text balances between the performance requirements presumed or imposed, their prediction during the design stage and the technology needed to realize the quality demanded.

Performance requirements discussed in 'Applied Building Physics: Boundary Conditions, Building Performance and Material Properties', stress the need for an excellent thermal insulation in cold and cool climates and the importance of a correct air, vapour and water management. It is therefore logical that Chapter 2 starts with a detailed overview of insulation materials, waterproof layers, vapour retarders, airflow retarders and joint caulking, after Chapter 1 recaptured the performance array at the building assembly level. In the chapters that follow the building assemblies that together shape a building are analyzed: foundations, basements and floors on grade, the load bearing structure, floors and massive facade systems. Each time the impact of the performance requirements on design and construction is highlighted. For decades, the Laboratory of Building Physics at the K. U. Leuven also did extended testing on highly insulated massive facade assemblies. The results are used and commented.

0.2 Units and symbols

The book uses the SI-system (internationally mandatory since 1977). Base units are the meter (m), the kilogram (kg), the second (s), the Kelvin (K), the ampere (A) and the candela. Derived units, which are important, are:

Unit of force: Newton (N); $1\,N = 1\,kg \cdot m \cdot s^{-2}$
Unit of pressure: Pascal (Pa); $1\,Pa = 1\,N/m^2 = 1\,kg \cdot m^{-1} \cdot s^{-2}$
Unit of energy: Joule (J); $1\,J = 1\,N \cdot m = 1\,kg \cdot m^2 \cdot s^{-2}$
Unit of power: Watt (W); $1\,W = 1\,J \cdot s^{-1} = 1\,kg \cdot m^2 \cdot s^{-3}$

For the symbols, the ISO-standards (International Standardization Organization) are followed. If a quantity is not included in these standards, the CIB-W40 recommendations (International Council for Building Research, Studies and Documentation, Working Group 'Heat and Moisture

Performance Based Building Design 1. From Below Grade Construction to Cavity Walls.
First edition. Hugo Hens.
© 2012 Ernst & Sohn GmbH & Co. KG. Published 2012 by Ernst & Sohn GmbH & Co. KG

Transfer in Buildings') and the list edited by Annex 24 of the IEA, ECBCS (International Energy Agency, Executive Committee on Energy Conservation in Buildings and Community Systems) are applied.

Table 0.1. List with symbols and quantities.

Symbol	Meaning	Units
a	Acceleration	m/s^2
a	Thermal diffusivity	m^2/s
b	Thermal effusivity	W/(m$^2 \cdot$ K $\cdot$ s$^{0.5}$)
c	Specific heat capacity	J/(kg $\cdot$ K)
c	Concentration	kg/m^3, g/m^3
e	Emissivity	–
f	Specific free energy	J/kg
	Temperature ratio	–
g	Specific free enthalpy	J/kg
g	Acceleration by gravity	m/s^2
g	Mass flow rate, mass flux	kg/(m$^2 \cdot$ s)
h	Height	m
h	Specific enthalpy	J/kg
h	Surface film coefficient for heat transfer	W/(m$^2 \cdot$ K)
k	Mass related permeability (mass may be moisture, air, salt …)	s
l	Length	m
l	Specific enthalpy of evaporation or melting	J/kg
m	Mass	kg
n	Ventilation rate	s^{-1}, h^{-1}
p	Partial pressure	Pa
q	Heat flow rate, heat flux	W/m^2
r	Radius	m
s	Specific entropy	J/(kg $\cdot$ K)
t	Time	s
u	Specific latent energy	J/kg
v	Velocity	m/s
w	Moisture content	kg/m^3
x, y, z	Cartesian co-ordinates	m
A	Water sorption coefficient	kg/(m$^2 \cdot$ s$^{0.5}$)
A	Area	m^2
B	Water penetration coefficient	m/s$^{0.5}$
D	Diffusion coefficient	m^2/s

0.2 Units and symbols

Symbol	Meaning	Units
D	Moisture diffusivity	m²/s
E	Irradiation	W/m²
F	Free energy	J
G	Free enthalpy	J
G	Mass flow (mass = vapour, water, air, salt)	kg/s
H	Enthalpy	J
I	Radiation intensity	J/rad
K	Thermal moisture diffusion coefficient	kg/(m·s·K)
K	Mass permeance	s/m
K	Force	N
L	Luminosity	W/m²
M	Emittance	W/m²
N	Vapour diffusion constant	s⁻¹
P	Power	W
P	Thermal permeance	W/(m²·K)
P	Total pressure	Pa
Q	Heat	J
R	Thermal resistance	m²·K/W
R	Gas constant	J/(kg·K)
S	Entropy, saturation degree	J/K, –
T	Absolute temperature	K
T	Period (of a vibration or a wave)	s, days, etc.
U	Latent energy	J
U	Thermal transmittance	W/(m²·K)
V	Volume	m³
W	Air resistance	m/s
X	Moisture ratio	kg/kg
Z	Diffusion resistance	m/s
α	Thermal expansion coefficient	K⁻¹
α	Absorptivity	–
β	Surface film coefficient for diffusion	s/m
β	Volumetric thermal expansion coefficient	K⁻¹
δ	Vapour conductivity	s
η	Dynamic viscosity	N·s/m²
θ	Temperature	°C
λ	Thermal conductivity	W/(m·K)

Symbol	Meaning	Units
μ	Vapour resistance factor	–
ν	Kinematic viscosity	m²/s
ρ	Density	kg/m³
ρ	Reflectivity	–
σ	Surface tension	N/m
τ	Transmissivity	–
ϕ	Relative humidity	–
α, ϕ, Θ	Angle	rad
ξ	Specific moisture ratio	kg/kg per unit of moisture potential
Ψ	Porosity	–
ψ	Volumetric moisture ratio	m³/m³
Φ	Heat flow	W

Table 0.2. List with suffixes and notations.

Symbol	Meaning
Indices	
A	Air
c	Capillary, convection
e	Outside, outdoors
h	Hygroscopic
i	Inside, indoors
cr	Critical
CO_2, SO_2	Chemical symbol for gasses
m	Moisture, maximal
r	Radiant, radiation
sat	Saturation
s	Surface, area, suction
rs	Resulting
v	Water vapour
w	Water
ϕ	Relative humidity
Notation	
[ ], bold	Matrix, array, value of a complex number
Dash	Vector (ex.: $\bar{a}$)

0.3 References and literature

[0.1] CIB-W40 (1975). Quantities, Symbols and Units for the description of heat and moisture transfer in Buildings: Conversion factors. IBBC-TNP, Report No. BI-75-59/03.8.12, Rijswijk.

[0.2] ISO-BIN (1985). Standards series X02-101 – X023-113.

[0.3] Kumaran, K. (1996). *Task 3: Material Properties.* Final Report IEA EXCO ECBCS Annex 24. ACCO, Leuven, pp. 135.

1 Performances

1.1 In general

This chapter starts by providing some definitions and the performance arrays. It then gives an analysis of the interaction between a rigorous application of performance metrics and building, followed by the possible impact of performance formulation on the construction process.

1.2 Definitions and basic characteristics

The term 'performance' encompasses all building-related physical properties and qualities that are predictable during the design stage and controllable during and after construction. Typical for performances is their hierarchical structure with the built environment as highest level (level 0) followed by the building (level 1), the building assemblies (level 2) and finally layers and materials (level 3). Relation between the four levels is typically top-down. 'Predictable' demands calculation tools and physical models that allow evaluating a design, whereas 'controllable' presumes the existence of measuring methods available on site. In some countries, the selection of building performance requirements had legal status. That coupled with a well-balanced enforcement policy guarantees application. One could speak of must and may requirements. Must is legally required, whereas may is left to the principal.

1.3 Advantages

The main advantage of a performance-based rationale is the objectification of expected and delivered building quality. For too long a time, designers juggled with 'the art of construction' without defining what kind of art was involved. With a rigorous application of performance metrics, the principal knows the physical qualities he may expect. In forensic cases, performance requirements provide a correct reference, which is not the case with the art of construction. A performance approach may also stimulate system based manufacturing. And finally, performance metrics could steer the building sector in a more research based direction.

1.4 Performance arrays

The basis for a system of performance arrays are the functional demands, the needs for accessibility, safety, well-being, durability, energy efficiency and sustainability and the requirements imposed by the usage of a building. For the arrays, see Table 1.1 and 1.2.

Table 1.1. Performance array at the building level (level 1).

Field		Performances
Functionality		Safety when used
		Adapted to usage
Structural adequacy		Global stability
		Strength and stiffness against vertical loads
		Strength and stiffness against horizontal loads
		Dynamic response
Building physics	Heat, air, moisture	Thermal comfort in winter
		Thermal comfort in summer
		Moisture tolerance (mould, dust mites, etc.)
		Indoor air quality
		Energy efficiency
	Sound	Acoustical comfort
		Room acoustics
		Overall sound insulation (more specific: flanking transmission)
	Light	Visual comfort
		Day-lighting
		Energy efficient artificial lighting
	Fire safety[1]	Fire containment
		Means for active fire fighting
		Escape routes
Durability		Functional service life
		Economic service life
		Technical service life
Maintenance		Accessibility
Costs		Total and net present value, life cycle costs
Sustainability		Whole building life cycle assessment and evaluation

[1] In countries like The Netherlands, Germany and Austria fire safety belongs to building physics. In other countries, it doesn't.

1.4 Performance arrays

Table 1.2. Performance array at the building assembly level (level 2).

Field		Performances
Structural adequacy		Strength and stiffness against vertical loads Strength and stiffness against horizontal loads Dynamic response
Building physics	Heat, air, moisture	Air-tightness • Inflow, outflow • Venting • Wind washing • Indoor air venting • Indoor air washing • Air looping
		Thermal insulation • Thermal transmittance (U) • Thermal bridging (linear and local thermal transmittance) • Thermal transmittance of doors and windows • Mean thermal transmittance of the envelope
		Transient response • Dynamic thermal resistance, temperature damping and admittance • Solar transmittance • Glass percentage in the envelope
		Moisture tolerance • Building moisture and dry-ability • Rain-tightness • Rising damp • Hygroscopic loading • Surface condensation • Interstitial condensation
		Thermal bridging • Temperature factor
		Others (i.e. the contact coefficient)
	Acoustics	Sound attenuation factor and sound insulation Sound insulation of the envelope against noise from outside Flanking sound transmission Sound absorption
	Lighting	Light transmittance of the transparent parts Glass percentage in the envelope
	Fire safety[1]	Fire reaction of the materials used Fire resistance
Durability		Resistance against physical attack (mechanical loads, moisture, temperature, frost, UV-radiation, etc.) Resistance against chemical attack Resistance against biological attack
Maintenance		Resistance against soiling Easiness of cleaning
Costs		Total and net present value
Sustainability		Life cycle analysis profiles

[1] In countries like The Netherlands, Germany and Austria fire safety belongs to building physics. In other countries, it doesn't.

1.5 Design based on performance metrics

1.5.1 The design process

'Designing' is multiply undefined. At the start, information is only indefinitely known. Each design activity may produce multiple answers, some better than others, which however cannot be classified as wrong. That indefiniteness demands a cyclic approach, starting with global choices based on sparse sets of known data, for buildings listed as project requirements and design intents. The choices depend on the knowledge, experience and creativity of the designer. The outcomes are one or more sketch designs, which then are evaluated based on the sets of imposed or demanded level 0 and 1 performance requirements. One of the sketch designs is finally optimized and the rest not meeting the performances are discarded. The result is a pre-design with form and spatiality fixed but the building fabric still open for adaptation.

With the pre-design, the set of agreed-on data increases. During the stages that follow, refinement alternates with calculations that have a double intent: finding 'correct' answers and adjusting the fabric to comply with the performance requirements imposed. That last phase ends with the final design, encompassing the specifications and the construction drawings needed to realize the building.

1.5.2 Integrating a performance analysis

Designing evolves from the whole to the parts and from vaguely to precisely known data and parameters. These are generated by the design itself, allowing performance analysis to become more refined as the design advances.

During the sketch design phase only level 1 performance requirements such as structural integrity, energy efficiency, comfort and costs receive attention. As most data are only vaguely known, only simple models facilitating global parametric analysis can be used. This isn't unimportant as decisions taken during sketch design fix many qualities of the final design.

At the pre-design stage along with level 1, the level 2 performance requirements also have to be considered as these govern translation of form and spatiality into building construction. As more parameters and data are established, evaluation can be more refined. The load bearing system gets its final form, the enclosure is designed and the first finishing choices are made. Options are considered and adjusted from a structural, building physical, safety, durability, maintainability, cost and sustainability point of view.

Detailing starts with the final design. Designing becomes analyzing, calculating, comparing, correcting and deciding about materials, layer thicknesses, beam, column and wall dimensions, reinforcement bars and so on. The performance metrics now fully operate as a quality reference. Proposed structural solutions and details must comply with all level 2 and 3 requirements, if needed with feedback to level 1. That way, performances get translated into solutions. Performances in fact do not allow construction. For that, each design idea has to be transformed into materials, dimensions, assemblies, junctions, fits, building sequences and buildability, with risk, reliability and redundancy as important aspects.

Performance requirements also should become part of the specifications, so contractors may propose alternatives on condition they perform equally or better for the same or lower price.

1.6 Impact on the building process

For decades, the triad <principal/architect/contractor> dominated the building process. The principal formulated a demand based on a list of requirements and intents. He engaged an architectural firm, which produced the design, all construction drawings with consultant's help (structural engineers, mechanical engineers and others), and the specifications on which contractors had to bid. The lowest bidder got the contract and constructed the building under supervision of the architect.

That triad suffers from drawbacks. The architect is saddled with duties for which he or she is hardly qualified. Producing construction drawings is typically a building engineering activity. Of course, knowledge about soil mechanics, foundation techniques, structural mechanics, building physics, building materials, building technology, and building services was procured but always after the pre-design was finished, that means after all influential decisions had been made. The split between design and construction further prevented buildability from being translated into sound construction drawings, which today, still, hardly differ sometimes from the pre-design ones. Details and buildability are left to the contractor, who may lack the education, motivation and resources for that. The consequences can be imagined. No industrial activity experiences as many damage cases as the building sector.

A performance rationale allows turning the triangle into a demand/bidder model. The demand comes from the principal. He produces a document containing the project requirements and intents. That document is much broader than a list of physical performances. Site planning, functional requirements at building and room level, form, architectural expression and spatiality are all part of it. Based on that document, an integrated building team, which includes the architect, all consulting engineers and sometimes the contractor is selected based on the sketch design it proposes. The assigned team has to produce the pre- and final drawings, included structure, building services, all energy efficiency aspects and, if demanded, an evaluation according to LEED, BREEAM or any other rating systems. If the contractor is part of the team, the assigned team also has to construct and decommission the building. Otherwise, a contractor is chosen based on a price to quality evaluation.

1.7 References and literature

[1.1] VROM (1991). *Teksteditie van het besluit 680* (Text edition of the decree 680). Bouwbesluit, Den Haag (in Dutch).

[1.2] Rijksgebouwendienst (1995). *Werken met prestatiecontracten bij vastgoedontwikkeling, Handboek* (Using performance based contracts for real estate development, handbook). VROM publicatie 8839/138, 88 p. (in Dutch).

[1.3] Stichting Bouwresearch (1995). *Het prestatiebeginsel, begrippen en contracten* (The performance concept, notions and contracts). Rapport 348, 26 p. (in Dutch).

[1.4] Australian Building Codes Board News (1995). Performance BCA, 14 p.

[1.5] CERF (1996). *Assessing Global Research Needs*. CERF Report #96-5016 A.

[1.6] Lstiburek, J., Bomberg, M. (1996). *The Performance Linkage Approach to the Environmental Control of Buildings*. Part 1, Journal of Thermal Insulation and Building envelopes, Vol. 19, Jan. 1996, pp. 224–278.

[1.7] Lstiburek, J., Bomberg, M. (1996). *The Performance Linkage Approach to the Environmental Control of Buildings.* Part 2, Journal of Thermal Insulation and Building envelopes, Vol. 19, April 1996, pp. 386–402.

[1.8] Hens, H. (1996). *The performance concept, a way to innovation in construction.* Proceedings of the 3rd CIB-ASTM-ISO-RILEM Conference 'Applications of the Performance Concept in Building', Tel Aviv, December 9–12, p. 5-1 to 5-12.

[1.9] Hendriks, L., Hens, H. (2000). *Building Envelopes in a Hollistic Perspective.* Final report IEA-Annex 32, IBEPA, Task A, ACCO, Leuven, 101 p. + add.

[1.10] ANSI/ASHRAE/USGBC/IES (2009). *Standard 189.1 for the design of high-performance green buildings except low-rise residential buildings.*

[1.11] ANSI/ASHRAE/USGBC/IES (2010). *189.1 User's manual.*

[1.12] Hens, H. (2010). *Applied building physics, boundary conditions, performances, material properties.* Wilhelm Ernst und Sohn (a John Wiley Company), Berlin.

2 Materials

2.1 In general

The second chapter first reviews materials used for thermal insulation. It then considers vapour barriers, also called vapour control layers, and air barriers, more generally known as air control layers. The last part examines joints between building components.

2.2 Array of material properties

Each knowledge field evaluates materials according to their properties. The storage and transport of heat, moisture and air in and across materials is also quantified that way, with density ρ and porosity Ψ – the weight per unit volume of material and the volume taken in by the pores in a unit volume of material – as basic characteristics. Even the consequence of heat, air and moisture presence is described using properties, with some combinations of properties mirroring unique physical features, see Table 2.1.

Table 2.1. Array of thermal, hygric and air-related material properties.

	Heat	Moisture	Air
Storage	Specific heat capacity c Volumetric specific heat ρc	Specific moisture ratio ξ Specific moisture content $\rho \xi$	Specific air content c_a
Transport	Thermal conductivity λ Thermal resistance R Absorptivity α Emissivity e Reflectivity ρ	Vapour permeability δ Vapour resistance factor μ Diffusion thickness μd Moisture permeability k_m Thermal moisture permeability K_θ	Air permeability k_a Air permeance K_a
Combined	Thermal diffusivity a $(= \lambda/(\rho c))$ Thermal effusivity b $(= \sqrt{\rho c \lambda})$	Moisture diffusivity D_w Water absorption coefficient A	
Consequences	Thermal expansion coefficient α	Hygric expansion ε	

2.3 Thermal insulation materials

2.3.1 Introduction

Thermal insulation materials were developed in order to minimize heat transport. That requires reducing thermal conductivity (λ) to the utmost, an objective that could not be reached without knowing how heat is transferred across a porous material.

2.3.2 Apparent thermal conductivity

2.3.2.1 In general

The property 'thermal conductivity' stands for the ratio between the vector 'heat flow rate' somewhere in a material and the vector 'temperature gradient' there. In isotropic materials, that ratio is a scalar whereas in anisotropic materials it is a tensor with a value along the x-, y- and z-axis: λ_x, λ_y and λ_z. For those materials, Fourier's second law becomes:

$$\lambda_x \frac{\partial^2 \theta}{\partial x^2} + \lambda_y \frac{\partial^2 \theta}{\partial y^2} + \lambda_z \frac{\partial^2 \theta}{\partial z^2} = \rho\, c\, \frac{\partial \theta}{\partial t} \qquad (2.1)$$

But this definition does not apply for a highly porous insulation material. Their apparent thermal conductivity is described as the heat passing a 1 m³ large cube with adiabatic lateral surfaces per unit time for 1 K temperature difference between top and bottom face. That condition is met per m² in an infinitely vast, 1 meter thick layer with 1 °C difference between both isothermal faces. Measurement of the apparent thermal conductivity is based on that description. A material sample of thickness d meter is mounted between a warm and a cold plate. Once steady state is reached, the temperature difference ($\Delta\theta$) over and heat flow (Φ) across the central part of the sample is logged. When the test apparatus is wrapped adiabatically and the central area A is small compared to the sample area, heat flow develops perpendicularly to thickness and the apparent thermal conductivity becomes:

$$\lambda = \frac{\Phi\, d}{A\, \Delta\theta} \qquad (2.2)$$

2.3.2.2 Impact of the transport modes

Heat flow across a dry porous material combines four transport modes (Figure 2.1): (1) conduction along the matrix, (2) conduction in the pore gas, (3) convection in that gas and (4) radiation in all pores between the pore walls. If humid, two additional modes intervene: (5) conduction in the adsorbed water and (6) latent heat transfer.

Apparent thermal conductivity as measured is not a fixed material property but a characteristic whose value depends on factors directly linked to these transport modes.

(1) + (2) Conduction along the matrix and in the pore gas (λ_c)

If only these two intervened, the equivalent thermal conductivity should be:

$$\lambda_c = \lambda_M (1-\Psi) + \lambda_G \frac{2\Psi}{1+\Psi} \qquad (2.3)$$

2.3 Thermal insulation materials

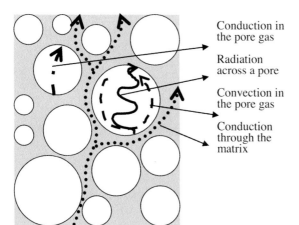

Figure 2.1. Heat transfer in a porous material.

where Ψ is total porosity, λ_M thermal conductivity of the matrix and λ_G thermal conductivity of the pore gas. According to that formula, apparent thermal conductivity lowers with increasing porosity. Porosity is now given by:

$$\Psi = \frac{\rho_s - \rho}{\rho_s} \qquad (2.4)$$

with ρ_s specific density of the matrix material. The same matrix material with higher porosity, yet with lower overall density, will thus see its apparent thermal conductivity drop, a fact proven experimentally (Figure 2.2). Also, a matrix material with lower thermal conductivity or a pore gas that insulates better than stagnant air gives relief. A very low apparent thermal conductivity is reached with vacuum pores ($\lambda_G \approx 0$).

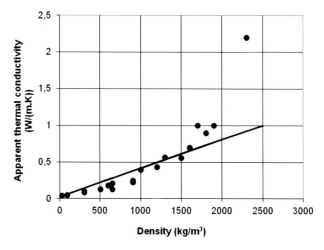

Figure 2.2. Apparent thermal conductivity versus density when only conduction in matrix and pore gas is present. The full line gives values for cellular glass, where glass forms the matrix, $\lambda_M = 1$ W/(m·K). The dots represent measured thermal conductivities for several materials.

But, when another gas than air fills the pores, diffusion of oxygen and nitrogen into the pores, diffusion of pore gas to the surroundings and adsorption of pore gas in the matrix material, slowly lifts the apparent thermal conductivity to a final equilibrium. The lift looks like this in the first weeks:

$$\lambda_c = \lambda_c(0) + C_1 \sqrt{t}$$

with coefficient C_1 inversely proportional to the diffusion resistance factor of the insulation material and proportional to the temperature with exponent n (T^n, $n < 1$, T in K). Later, lifting slows down to:

$$\lambda_c = \lambda_c(0) + \left[\lambda_c(\infty) - \lambda_c(0)\right]\left[1 - \exp(C_2 t)\right]$$

where coefficient C_2 depends on diffusion resistance factor and temperature as C_1 does. A high diffusion resistance factor means a slow lift. Or, if one wants to store another gas than air in the pores, the matrix should be as vapour retarding as possible. An alternative is to face the insulation boards with a vapour-tight lining.

Thus, an insulation material must be low density; the pores should be filled with an insulating gas that is better than stagnant air and have a matrix that conducts heat inefficiently.

(3) Convection in the pores

Convection only develops in larger pores. Its impact is quantified by multiplying the thermal conductivity of the pore gas in Equation (2.3) with the Nusselt number ($X \geq 1$):

$$\lambda_c = \lambda_M (1 - \Psi) + X \lambda_G \frac{2\Psi}{1+\Psi} \tag{2.5}$$

Convection increases heat flow. Or, pores in an insulation material should be so small that the Nusselt number is 1. A good insulation material thus is not only very porous; the pore volume must consist of very small pores.

(4) Radiation in the pores

In every pore, except if perfectly reflecting, pore walls at different temperatures exchange radiant heat. The result is a radiant term (λ_R) complementing the apparent thermal conductivity of formula [2.5]:

$$\lambda = \lambda_M (1-\Psi) + X \lambda_G \frac{2\Psi}{1+\Psi} + \underbrace{\left\{ F_{RC} \frac{4 C_b T_m^3 d}{100^4 \left(\dfrac{1}{e_1} + \dfrac{1}{e_2} + n \dfrac{1+\rho-\tau}{1-\rho+\tau} - 1\right)} \right\}}_{\lambda_R} \tag{2.6}$$

with:

$$F_{RC} = 1 + 100^4 \frac{\lambda_c (1/e_1 + 1/e_2 - 1)}{4 C_b T_m^3 d} \left[\frac{\Delta\theta_1 + \Delta\theta_n}{\Delta\theta/n} - 1\right] \tag{2.7}$$

2.3 Thermal insulation materials

In these formulas, n represents the number of pore walls along the insulation thickness d. ρ and τ are long wave reflectivity, respectively long wave transmissivity of the pore walls, while e_1 and e_2 stand for the long wave emissivity, side insulation, of the linings at both sides. F_{RC} is a correction factor, accounting for the interaction between radiation, convection and conduction. In it, $\Delta\theta$ is the temperature difference across the thickness d of the insulation material, whereas $\Delta\theta_1$ and $\Delta\theta_n$ are the temperature differences between the facing at both sides and the first pore wall encountered.

The radiant term has a 'temperature to the 3th power' impact on the apparent thermal conductivity. As the thermal conductivity of matrix and pore gas is also temperature sensitive, the overall dependence is:

$$\lambda = \lambda_0 + a_1\, \theta^n + a_2\, \theta^3 \tag{2.8}$$

with $0 < n < 1$. For $-20 \leq \theta \leq 50$ °C, [2.8] is closely matched by:

$$\lambda = \lambda_0 + a_R\, \theta = \lambda_0\left(1 + a'_R\, \theta\right)$$

The lighter the insulation material, i.e. the thinner the pore walls or the larger the pores, the higher the coefficient a_R and the more temperature dependant the apparent thermal conductivity is. As there are less large pores across the layer thickness than small pores, whereas thin pore walls have higher long wave transmissivity than thicker pore walls, in both cases, radiation gains importance.

A radiant side effect is an apparent thermal conductivity increase with insulation thickness. In fact, as the number of pore walls n can be replaced by the ratio between layer thickness d and mean pore width d_P (d/d_P), layer thickness is explicitly present in the numerator and hidden in the denominator of Equation (2.6). That way, the term shifts from zero for layer thickness zero to an asymptote $\lambda_{R\infty}$ for infinite thickness:

$$\lambda_{R\infty} \approx \frac{4\, F\, C_b\, T_m^3\, d_P}{100^4 \left[\dfrac{1+\rho-\tau}{1-\rho+\tau}\right]} \tag{2.9}$$

This asymptote increases with larger pores ($d_P\uparrow$) and the pore walls transmitting more radiation, i.e. when an insulation material has a lower density. That way, radiant exchanges obstruct apparent thermal conductivity from a continuous drop with density. In fact, once below a limit density, a further drop is only possible by enlarging the pores or thinning the pore walls. In both cases, radiation gains importance, turning the apparent thermal conductivity into a sum of a monotonously decreasing conductive and increasing radiant part. That way, an optimum density exists for which at a given layer thickness the equivalent thermal conductivity is minimal. As a formula:

$$\lambda = b_1 + b_2\, \rho + b_3/\rho \tag{2.10}$$

for mineral wool at 20 °C (see Figure 2.3):
$b_1 = 0.039$ W/(m · K), $b_2 = 4.4 \cdot 10^{-3}$ W · m²/(K · kg) and $b_3 = 0.289$ W · kg/(m⁴ · K)

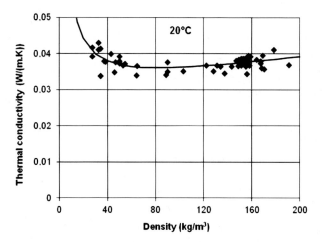

Figure 2.3. Mineral wool: apparent thermal conductivity versus density.

(5), (6) Conduction in the adsorbed water layers, latent heat exchanges

Moist materials not only see extra heat conduction in the adsorbed water layers and condensed water islands, they also suffer from latent heat flow by evaporation/diffusion/condensation/backflow in the pores. That extra conduction results in a linear relationship between apparent thermal conductivity and moisture ratio for porous building materials and a parabolic relationship between apparent thermal conductivity and volumetric moisture ratio for insulation materials:

$$\lambda = \lambda_d \left(1 + a_X \, X\right) \qquad \lambda = \lambda_d \left(1 + a_\Psi \, \Psi + b_\Psi \, \Psi^2\right) \qquad (2.11)$$

where λ_d in both equations is the apparent thermal conductivity for the dry material. Latent heat flow however adds a term:

$$q_v = l_b \, g_v = -\frac{l_b}{\mu \, N} \frac{dp_{sat}}{d\theta} \, \text{grad} \, \theta \qquad (2.12)$$

where l_b is the latent heat of evaporation. Apparent thermal conductivity thus becomes:

$$\lambda = \lambda_d \left(1 + a_X \, X\right) \left[1 + \frac{l_b}{\mu \, N \, \lambda_d \left(1 + a_X \, X\right)} \frac{dp_{sat}}{d\theta}\right]$$

$$\approx \lambda_d \left(1 + a_X \, X\right) \left[1 + \frac{4.6 \cdot 10^{-4} \, p_{sat}}{\mu \, \lambda_d \left(1 + a_X \, X\right) T} \left(\frac{7066.27}{T} - 5.976\right)\right] \qquad (2.13)$$

Thanks to evaporation/diffusion/condensation/backflow in the pores, temperature affects apparent thermal conductivity in humid insulation materials more than by radiation only. Influence also quickly increases with a decreasing vapour resistance factor (μ). The largest impact in fact is seen in vapour permeable materials such as mineral wool, where evaporation of moisture at the warm side causes a jump in apparent thermal conductivity (Figure 2.4).

2.3 Thermal insulation materials

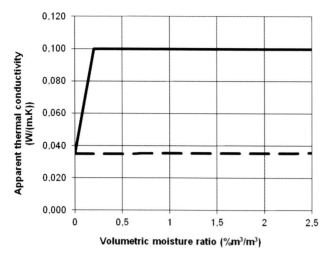

Figure 2.4. Mineral wool: apparent thermal conductivity versus volumetric moisture ratio. The unbroken line is moisture at the warm side; the dotted line represents moisture at the cold side.

2.3.3 Other properties

2.3.3.1 Mechanical

Due to their very high porosity, insulation materials have limited strength and stiffness. Under mechanical load, they behave like mattresses rather than elastic-plastic. Low stiffness in turn incurs creep, relaxation and sometimes remarkable form instabilities. Therefore insulation materials hardly may perform load bearing functions, although there are exceptions like good compression strength when used in floors and foundations whereas in sandwich panels the insulation layer must withstand shear.

2.3.3.2 Physical

Moisture

Most insulation materials are non-hygroscopic. All in fact have macro-pores and limited specific pore surface. Consequently, they hardly adsorb water vapour whereas capillary condensation only happens at relative humidity near 100%. However, excluding capillary action in fibrous insulation materials requires water-repellent treatment. Only closed-pore insulation materials guarantee imperviousness for water heads, while limiting vapour diffusion across the pores and interstitial condensation in the pores demands a high vapour resistance factor, again requiring closed pores. That favours foams as opposed to fibrous materials, which are vapour permeable, pervious for water heads and non-capillary only when treated with a hydrophobic resin. Whether insulation materials lose strength and stiffness, degrade biologically and rot when moist, depends on the matrix material.

Air

Good air-tightness demands closed pores. Foams are no problem, but not with fibrous materials, which are extremely air permeable.

2.3.3.3 Fire

Also here the matrix material qualifies. Insulations based on organic and synthetic materials typically belong to the classes 'flammable' whereas the non-organic ones are mostly inflammable.

2.3.3.4 Sensitivity to temperature, IR and UV

Again, the matrix material plays the main role. Synthetics behave worse whereas organic and non-organic materials hardly give problems.

2.3.4 Materials

Insulating building materials, insulation materials and new developments are discussed using following scheme.

Short description
Properties   Density
                      Heat
                      Moisture
                      Air
                      Strength and stiffness
Behaviour   In general
                      Under mechanical load
                      Sensitivity to temperature, IR and UV
                      Under moisture load
                      Exposure to fire
                      Others
Usage

In addition, some attention is given to radiant barriers.

2.3.4.1 Insulating building materials

Brick masonry

Increasing the thermal resistance of brick masonry demands (1) limiting the meters run of horizontal and head joints per m² of wall, (2) developing lower density bricks, (3) using insulating mortars and (4) increasing wall thickness. Less meter run of joints means using fast bricks. Lower density combines lighter potsherd with optimal perforation patterns.

Larger format   After world war II the massive brick, $L \times W \times H = 19 \times 9 \times 6.5$ cm, has been replaced by fast bricks, $29 \times 14 \times 14$ or $29 \times 19 \times 14$ cm.
Lighter potsherd   Possible by mixing sawdust or polystyrene pearls in the clay. During firing, sawdust carbonizes and polystyrene sublimates. The result is a cloud of macro pores in the fired brick, which lowers density from 1800 kg/m³ down to less than 1000 kg/m³.
Optimal perforation   Perpendicular or diagonal perforations extend transmission path (Figure 2.5).

2.3 Thermal insulation materials

Insulating mortar — Produced by replacing sand by a perlite or vermiculite granules fraction.

Figure 2.5. Perforation patterns in fast bricks: (2) and (3) better than (1).

Properties

Density — Lies between 750 and 880 kg/m³ for lightweight fast bricks. Dense brickwork may weigh 2000 kg/m³.

Thermal

Specific heat capacity — Dry 840 J/(kg·K), independent of density

Thermal resistance — Up to 0.5 m²·K/W for a 14 cm thick lightweight fast brick wall, density 900 kg/m³. For comparison, a 14 cm thick normal fast brick wall does not pass 0.28 m²·K/W. A 30 cm thick light weight fast brick wall could reach 1.7 m²·K/W.

Hygric

Moisture content — Bricks are hardly hygroscopic. They have high capillary moisture content.

Diffusion thickness — Masonry has a lower diffusion thickness than the bricks due to badly filled joints and micro cracks between bricks and joints. A good estimate is $\mu_{eq} d \approx 5 d$ (m).

Capillary water absorption coefficient — From moderate (0.05 kg/(m²·s$^{0.5}$)) to high (0.8 kg/(m²·s$^{0.5}$)), depending on the brick's porous structure.

Usage

Lightweight fast brickwork is well suited as inside leaf in cavity walls and for massive walls with rain right outside render. However, it does not replace a good thermal insulation. For that, the apparent thermal conductivity is too high. In addition, embodied energy is not negligible.

Concrete

The first step on the way to a better insulating concrete consists of replacing gravel by lighter particles: furnace slag, expanded clay, perlite or polystyrene pearls. Density and apparent thermal conductivity drop as does strength and stiffness, but shrinkage and creep increase. The lowest apparent thermal conductivity is attained by skipping gravel or any other addition and foaming the sand/mortar mixture through gas formation in an autoclave. The result is 'aerated concrete', manufactured in blocks with dimensions up to 59.5 × 29.5 × 29.5 cm and as facade or roof elements. Its foamed structure allows sawing and milling.

Properties

Density
While 'normal' concrete weights ≈ 2200 kg/m^3, expanded clay concrete has a density between 650 and 1600 kg/m^3, with 1600 kg/m^3 for structural application and 650 kg/m^3 for non-bearing uses. Polystyrene concrete weights 260–800 kg/m^3, 260 kg/m^3 as post-fill. Aerated concrete ranges between 350 and 800 kg/m^3.

Thermal
Specific heat capacity
Dry 840 J/(kg · K), independent of density

Apparent thermal conductivity
Tables attribute 1.6 to 2 W/(m · K) as 'dry value' to 'normal' concrete. Measurement gave 2.6 W/(m · K). Expanded clay concrete gives 0.024 exp (0.0027 ρ) for 600 < ρ < 1200 kg/m^3. With polystyrene concrete, the value is 0.041 exp (0.00232 ρ) for 250 < ρ < 800 kg/m^3. For aerated concrete it is: 0.12 + 0.000375 ρ for 450 < ρ < 620 kg/m^3.

Hygric
Moisture content
The cement gel turns concrete into a hygroscopic material.

Diffusion thickness
Drops with decreasing density and increasing moisture content.

Capillary water absorption coefficient
Concrete is not very capillary. First the second digit behind the decimal point differs from zero.

Strength and stiffness
Although density is quite low, constructing 4–5 floors high is still possible with heavier aerated concrete blocks.

Behaviour

Under moisture load
Hygric shrinkage increases with lower density. Reason is less particle resistance when going from gravel over expanded clay and polystyrene pearls to no particle resistance at all with aerated concrete! When building with that material, shrinkage demands proper detailing. Aerated concrete also has high building moisture content, up to 200–250 kg/m^3 and an initial thermal conductivity, which exceeds the air-dry value of 0.14–0.2 W/(m · K).

Exposure to fire
Due to the combination of non-combustibility, good insulation and low thermal expansion, aerated concrete has excellent fire resistance.

Usage

Aerated concrete is an alternative to thermal insulation. An air-dry aerated concrete wall, thickness 30 cm, has a clear wall thermal transmittance below 0.5 W/(m^2 · K).

2.3 Thermal insulation materials

2.3.4.2 Insulation materials

A material is called 'insulating', when its dry apparent thermal conductivity does not pass 0.07 W/(m · K). Classification happens according to structure, behaviour, application or matrix material. In the case of matrix material, the scheme becomes:

Group	Material	Acronym
Organic isolation materials	Cork	K
	Cellulose fibre	C
	Sea grass, wool, straw, flax	
Inorganic isolation materials	Glass fibre	MW
	Mineral wool	MW
	Cellular glass	CG
	Perlite, vermiculite	
Plastic foams	Expanded polystyrene	EPS
	Extruded polystyrene	XPS
	Polyurethane foam	PUR
	Polyisocyanurate foam	PIR
	Phenol, ureumformaldehyde and polyethylene foam	
Mixed materials	Pressed perlite boards	PPB

Only the materials in standard letters are commented in detail.
We also discuss new developments such as transparent (TIM) and vacuum insulation (VIP).

Sea grass, wool, straw and flax are called 'sustainable' by bio-ecologists, only because they are 'natural'. Their quality in terms of 'durability' however, is never referred to. All four are hygroscopic, moisture sensitive and combustible. Many people are allergic to wool. Upgrading durability and lowering combustibility demands addition of chemicals, among them borax salts. Whether these materials are still 'natural' with these additions is left unmentioned.

Cork

The basic material is the stripped bark of the cork oak. After grinding, the bark particles are autoclaved in steam at 350 °C. That expands them, kills moulds and bacteria, while part of the VOCs evaporates and the resin binds the particles into blocs. These are then cut to size. An alternative is to dry heat the cork particles, drench them into bitumen and press that mixture into boards.

Properties

Density	Between 80 and 250 kg/m³. Quite high for an insulation material.
Thermal	
Specific heat capacity	Dry ±1880 J/(kg · K), independent of density
Apparent thermal conductivity	For a density of 111 kg/m³, temperature between 0 and 0 °C and volumetric moisture ratio between 0 and 6% m³/m³:

$$\lambda = 0.042\left(1+1.8 \cdot 10^{-3}\,\theta\right) \quad \lambda = 0.042\left(1+4.3 \cdot 10^{-2}\,\Psi\right)$$

Hygric

Moisture content	Due to its organic origin, cork is hygroscopic. It also shows some capillarity.
Vapour resistance factor	Between 5 and 20. The value drops with higher relative humidity (in reality with moisture content).
Air	Its open porous structure makes cork air permeable.
Strength and stiffness	Cork has low compressive strength. A 0.11 MPa large compression during one day results in 10% strain for 145 kg/m³ dense boards.

Behaviour

Under mechanical load	Cork creeps. 1 day at 0.05 MPa compression increases strain from 1.5 to 5%. Allowable stress is therefore limited to 1/3 of that at 10% strain (σ_{10}).
Sensitivity to temperatures, IR and UV	Cork scores quite well. Thermal expansion coefficient is quite high ($\pm 40 \cdot 10^{-6}$ K^{-1}), but resistance against low and high temperatures is excellent and UV gives some discoloration only.
Moisture load	Like all organic materials, cork swells when wetted and shrinks when drying. If wet for a long enough period, it turns mouldy and may rot.
Exposure to fire	Cork burns.

Usage

Although cork was well suited to insulate low-sloped roofs and refrigerators, plastic foams took over that segment of the market. Never apply cork where high relative humidity is likely! Using it to upgrade airborne and contact sound insulation also makes no sense, as the material is too stiff. However heavy boards do well as vibration damper.

Cellulose

The basic materials are unused newspapers. To limit combustibility and mould sensitivity, the fibres are mixed with borax salts. The material applies as dry or wet sprayed loose fill. It is also available as dense boards.

Properties

Density	Ranges from 24 to 60 kg/m³. Depends among others on spraying pressure.
Thermal	
Specific heat capacity	Dry $\approx$ 1880 J/(kg · K), independent of density
Apparent thermal conductivity	Air dry given by

$$\frac{d/1000\,(1 + 0.00289\,(\theta - 24))}{(0.205 + 0.0247\,d) - (0.00201 + 0.0000143\,d)\,\rho}$$

with d thickness in mm, θ temperature in °C and ρ density in kg/m³

2.3 Thermal insulation materials

Hygric

Moisture content — Due to their organic origin, cellulose fibres are hygroscopic. The borax salts still increase sorption, see Figure 2.6. The fibres also show capillarity.

Vapour resistance factor — Is not higher than 1.9 for a density of 50 kg/m³. Drops with increasing moisture content. The low value is due to the fibrous structure.

Air — Fibrous structure makes the material air permeable, $k_a \approx 1.6 \cdot 10^{-3}$ s.

Strength and stiffness — Loading loose fill beyond the weight of the fibrous mass is excluded.

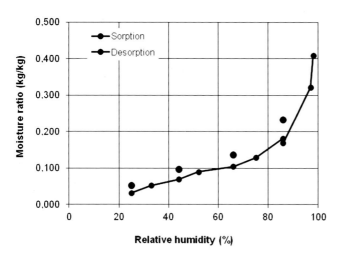

Figure 2.6. Cellulose fibre: sorption/desorption.

Behaviour

Under mechanical load At low density, static and dynamic forces induce irreversible settling (s, t in years). Measured (a for annum, year):

$$s = 100 \, \rho \left\{ 1/a + t/b + \left[1 - \exp(-dt) \right]/c \right\}$$

with	ρ kg/m³	a kg/m³	b a·kg/m³	c a·kg/m³	d a⁻¹
	30	1.50	6827.9	247.2	1.87
	35	1.75	8018.3	288.4	1.87
	40	2.00	9165.9	329.6	1.87

Under moisture load — Cellulose shows quite some hygric swelling and shrinkage. Wet spraying results in drying shrinkage. At moisture contents above 20% kg/kg, the fibres clog and may start rotting.

Exposure to fire — Despite borax salt addition, cellulose fibres are combustible. With thicker insulations, such as in passive houses, collapsing roofs during a fire create hazardous situations for fire fighters.

Drawbacks Cellulose dust may induce respiratory problems. During spraying, a mask should be worn. Also, the borax salts are not without problems. Simple exposure can cause respiratory and skin irritation. Ingestion of the salts may give gastrointestinal distress including nausea, persistent vomiting, abdominal pain, and diarrhoea. Effects on the vascular system and brain include headaches and lethargy, but are less frequent. In severe poisonings, a beefy red skin rash affecting palms, soles, buttocks and scrotum has been described. With severe poisoning, erythematous and exfoliative rash, unconsciousness, respiratory depression, and renal failure happen.

Usage

Cellulose fibres are an alternative to glass fibre and mineral wool. Typical applications are insulation of timber-framed walls, insulation of timber low-sloped roofs with insulation between purlins and insulation of attic floors. The dense boards may be used to insulate pitched roofs. However, cellulose fibres should never be used in air spaces exposed to very high relative humidity. This could be a problem when used in low-slope roofs and timber-framed walls with brick veneer. There solar driven vapour flow from wet veneers back to the inside during warm weather may humidify the fibres

Glass fibre and mineral wool

Glass fibre is produced using (recycled) glass, whereas mineral wool has diabase stone as a basic material. Glass and stone are melted, after which a spinning head stretches the melt into fibres with diameter < 10 µm. These fall through a spray of phenol or silicon binder on a conveyor belt, on which the facings for blankets and bats lie. Conveyor belt and fibre blankets, bats or boards then pass a heated press where the binder hardens and the insulation gets its final density and thickness. Then the blankets, bats and boards are cut to size. The spectrum of finished products ranges from loose fill over blankets and bats to soft, semi-dense and very dense boards.

At first sight, the two materials are similar. But there are differences. Glass fibre consists of well-ordered, long fibres whereas mineral wool is composed of unordered short fibres. Glass is amorphous, and diabase stone is crystalline.

Properties

Density For glass fibre 10 to 150 kg/m^3, for mineral wool 30 to 190 kg/m^3

Thermal

Specific heat capacity Same value as for stony materials, dry ±840 J/(kg · K)

Apparent thermal conductivity Glass fibre at 20 °C: $0.0262 + 5.6 \cdot 10^{-5} \rho + 0.184/\rho$.
Mineral wool at 20 °C: $0.0331 + 3.2 \cdot 10^{-5} \rho + 0.221/\rho$.

For both, the temperature impact is greatest at low density. Glass fibre clearly has a somewhat lower apparent thermal conductivity than mineral wool at the same density. Better production methods have resulted in further lowering these values to 0.032 W/(m · K).

2.3 Thermal insulation materials

Hygric

Moisture content — Glass fibre and mineral wool are hardly hygroscopic. Except when treated hydrophobically, the boards show some capillary action. High-density hydrophobic boards withstand a limited water head.

Vapour resistance factor — Very low: 1.2 to 1.5. Due to the fibrous structure.

Air — The fibrous structure makes glass fibre and mineral wool highly air permeable:
$k_a = 4.3 \cdot 10^{-3} \rho^{-1.3}$ respectively $2.1 \cdot 10^{-2} \rho^{-1.5}$ kg/(m·s·Pa).

Strength and stiffness — Blankets cannot take any load, except their own weight. Dense boards are moderately compression resistant ($\sigma_{10} \approx 0.04$–0.08 MPa).

Behaviour

Sensitivity to temperatures, IR and UV — Glass fibre and mineral wool are very temperature resisting. Thermal expansion coefficient is low ($\approx 7 \cdot 10^{-6}$ K^{-1}) and irreversible deformation under temperature load seldom occurs, though the binder may evaporate beyond 250 °C and degrade above 600 °C for glass fibre and 850 °C for mineral wool, which is therefore preferred for high temperature applications.

Under moisture load — Both are quite moisture tolerant, although wet blankets and bats lose their shape whereas wet dense boards lose stiffness and compression strength. Exposed to a combination of high temperature, moisture and oxygen, glass fibre slowly pulverizes.

Exposure to fire — Neither glass fibre nor mineral wool burn. The binder however may. Binder concentrations below 4% kg/kg evaporate, above it burns. Also, most facings are flammable.

Drawbacks — Wasps and rodents make nests in both materials.

Usage

Glass fibre and mineral wool are universal insulation materials. Applications range from low-slope roofs (dense boards) to pitched roofs (blankets, bats and soft boards), cavity fill (semi dense, water-repellent boards), timber frame insulation, EIFS (dense boards), floor insulation (dense boards) and perimeter insulation (dense boards). Manufacturers develop specific products for every application. There are boards with upgraded water repellence for full cavity fill, boards with very dense upper layer for low-slope roof application, etc.

Comment

In the nineties, there was some concern about the possible cancerous nature of mineral fibres. Where this is a fact for asbestos fibre, no proof was found for glass fibre and mineral wool. The fibres irritate skin and mucous membranes. During installation, wearing protective clothing and a mask is mandatory.

Cellular glass

The basic material is used glass bottles. These are melted and extracted as thin-walled pipes. After cooling, the pipes are ground and carbon dust added. That mixture is then poured in moulds that enter the furnace. While the glass melts, the carbon reacts explosively to form CO_2, giving a porous glass mixture that solidifies into cellular glass breads. These are cut into boards with dimensions 0.4×0.6 m² or 0.6×1 m² and faced if necessary.

Properties

Density	Between 100 and 500 kg/m³
Thermal	
Specific heat capacity	Same value as for stony materials, 840 J/(kg · K)
Apparent thermal conductivity	As a function of density, at 20 °C: $0.0405 + 1.9 \cdot 10^{-4} (\rho - 100)$. Impact of temperature for 129 kg/m³: $0.0464 (1 + 5.17 \cdot 10^{-3} \theta)$.
Hygric	
Moisture content	Cellular glass is not hygroscopic. Thanks to its closed pore structure, the material is neither capillary nor does it become wet under a water head.
Vapour resistance factor	Due to the closed pore structure, extremely high. Manufacturers claim unlimited. It is safer to say the number cannot be measured with a cup test. A value up to 70 000 is an acceptable guess.
Air	Cellular glass is airtight.
Strength and stiffness	Of all insulation materials, cellular glass has the highest compression strength ($\sigma_{10} \approx 0.5$ à 1 MPa), behaves elastically and is insensitive to creep. Despite this, loading is limited to 1/3 of σ_{10}. Care should be taken with local loads because the boards are brittle and have limited tensile strength.

Behaviour

Sensitivity to temperature, IR and UV	Cellular glass is very temperature tolerant. The thermal expansion coefficient is the same as for glass ($8 \cdot 10^{-6}$ K^{-1}). Irreversible deformation is excluded.
Under moisture load	Because cellular glass is water- and vapour tight, one should not expect any moisture attack. Frost nevertheless may cause problems. When cutting boards, the surface pores are transected. This way, they can fill with water. Freezing then lets the expanding ice crush the pore walls, which allows water to fill the pores below. That way, repetitive frost/thawing creates a progressing front of broken pores, allowing the boards to become wet.
Exposure to fire	Cellular glass does not burn, though it pulverizes when flamed.
Others	Most bases and acids do not attack the material.

2.3 Thermal insulation materials

Usage

Cellular glass is an expensive insulation material. Therefore, one should only apply it where its properties add value. Examples are: insulation below foundations, thermal cut material, insulation for parking decks and industrial floors, insulation in envelope assemblies where form stability is a plus or the high quality compensates lack of maintenance (low-sloped roofs) or insulation where resistance against bases and acids is a requirement (industrial applications).

Comment

Manufacturers heavily stress vapour tightness, air-tightness and insensitivity to water uptake. These statements should be placed in context. Whereas a cellular glass board is extremely vapour- and airtight, a layer of 0.4×0.6 m² large boards may not be. It suffices that joints between the boards are loose.

Expanded polystyrene (EPS)

The basic material is pentane blown polystyrene pearls. In a first step, the pearls are heated beyond 100 °C, a temperature at which the evaporating pentane causes expansion. The expanded pearls are then stored for a few days allowing diffusion of remaining pentane. Then they are poured in steel moulds and heated a second time, now with steam. As a result, the expanded pearls coagulate in their own melt. Once cooled, the blocks are cut into boards, which are then stored until completion of initial shrinkage. EPS is a thermoplastic.

Properties

Density	Between 15 and 45 kg/m³. Manufacturers used to mark the boards according to density: PS15, PS20, PS30, etc. That changed with the standard EN 13163. Now boards are marked according to their compression strength at a deformation of 10% (in MPa).
Thermal	
Specific heat capacity	≈ 1470 J/(kg·K)
Apparent thermal conductivity	With density at 20 °C: $0.0174 + 1.9 \cdot 10^{-4} \rho + 0.258/\rho$. For a density of 15 kg/m³ as a function of a temperature/volumetric moisture ratio: $0.0354 (1 + 4.52 \cdot 10^{-3} \theta) / 0.0393 (1 + 0.048 \Psi)$
Hygric	
Moisture content	EPS is hardly hygroscopic and non-capillary, though macro pores between the expanded pearls fill under water head.
Vapour resistance factor	$35 + 2.1 (\rho - 15)$ with ρ (density) > 15 kg/m³. Standard deviation: $\sigma_{35} = 14$, $\sigma_{2.1} = 0.25$. The value hardly depends on relative humidity.
Air	Due to its granular structure, not airtight.
Strength and stiffness	Compression resistance increases with density. At 20 °C, $\sigma_{10} \approx 0.08$ MPa for a density of 15 kg/m³, ≈ 0.12 MPa for a density of 20 kg/m³ and ≈ 0.22 MPa for a density of 30 kg/m³. These values decrease at higher temperature. $\sigma_{10}/3$ is allowable.

Behaviour

Under mechanical load As for all plastic foams, creep increases with temperature.

Sensitivity to temperature, IR and UV Being a thermoplastic, EPS not only has a high thermal expansion coefficient, $80 \cdot 10^{-6}$ K^{-1}, but also temperature tolerance is poor. Above 70 °C, pore anisotropy, air/water vapour overpressure and softening pore walls create irreversible deformation. Above 80 °C it melts.

Under moisture load Although EPS has a good moisture tolerance, water condensing in the pores increases form instability. In fact, when humid, vapour saturation pressure, which augments exponentially with temperature, pushes pore gas pressure to higher values.

Exposure to fire Fire response is bad. EPS melts, accelerating fire spread that way. Additives are used to make the foam self-extinguishing. Such boards melt slower, while producing a black fatty smoke.

Drawbacks Tar and volatile oil products dissolve EPS. Rodents and insects devour it. But, it is an effective electrical insulator.

Usage

EPS is well suited for applications where temperatures do not exceed 70 °C and protection against rodents is guaranteed: EIFS-systems, cavity insulation, floor insulation, pitched roof insulation, etc. Usage indoors is allowed if finished with fire resisting lining. EPS could be used in low-sloped roofs, on condition the boards have a bituminous glass fibre facing at both sides and the roofing membrane gets gravel ballast.

Comment

EPS does not contain CFC's. Some styrene may be released.

Extruded polystyrene (XPS)

The basic material is polystyrene pearls. These are melted and a propellant added, after which an Archimedes screw presses the expanding melt through an extruder, which gives the boards a dense skin. The extruded foam is then stabilized in a waterbed and cut into separate boards. As EPS, XPS is stored during some six weeks before application.

Properties

Density Between 25 and 45 kg/m³ depending on the kind of application

Thermal

Specific heat capacity ±1470 J/(kg · K)

Apparent thermal conductivity With density at 20 °C:
$0.0241 + 1.3 \cdot 10^{-4} \Psi + 5.9 \cdot 10^{-5} \Psi^2$.

With volumetric moisture ratio at 10 °C:
$0.0241 + 1.3 \cdot 10^{-4} \Psi + 5.9 \cdot 10^{-5} \Psi^2$.

Apparent thermal conductivity is very low. The reason is the propellant, first Freon 12, now HCFC's and others. Diffusion results in a slow increase of the apparent thermal conductivity over time to an equilibrium.

2.3 Thermal insulation materials

Hygric

Moisture content — XPS as closed pore foam is neither hygroscopic nor capillary and shows no moisture uptake under water head.

Vapour resistance factor — High thanks to the closed pores and the dense surface layer: $114 + 3.42 \, (\rho - 20)$ for a density beyond 20 kg/m^3.

Standard deviation:
$\sigma_{114} = 33$ and $\sigma_{3.42} = 0.61$.

The value does not depend on relative humidity but drops with board thickness d: $\mu = 160 + 0.451/d$ (d in m). The two dense surface layers created by extrusion in fact have a much higher vapour resistance number than the bulk of the boards. The thinner these are, the larger the impact of the two surface layers.

Air — XPS is airtight.

Strength and stiffness — Compression resistance increases with density. At 20 °C, $\sigma_{10} \approx 0.15$–0.25 MPa for 25 kg/m^3. With 30 kg/m^3 σ_{10} increases to 0.25–0.30 MPa. 35 kg/m^3 gives $\sigma_{10} \approx 0.5$ MPa and 45 kg/m^3 $\sigma_{10} \approx 0.6$–0.7 MPa.
These values decrease at higher temperature. $\sigma_{10}/3$ is allowable.

Behaviour

Under mechanical load — Creep increases with temperatures.

Sensitivity to temperature, IR and UV — As a thermoplastic, XPS not only has a high thermal expansion coefficient, $\pm 80 \cdot 10^{-6}$ K^{-1}, but heat resistance is also poor with irreversible deformation at temperatures above 70 °C. Above 80 °C, it melts. The causes of form instability are the same as for EPS. Because of a higher stiffness, the consequences are more troubling.

Under moisture load — Of all isolation materials, XPS has the highest moisture tolerance. Being non-hygroscopic, non-capillary and picking up no moisture under water head, the dense surface layers also exclude frost problems. Only interstitial condensation may wet the pores. This however is a very slow process thanks to the high vapour resistance factor. Nevertheless, when used between two moist layers, XPS-boards may get quite wet after ten to twenty years with higher apparent thermal conductivity as the main consequence.

Exposure to fire — Negative. Fire retarding XPS extinguishes when removing the flame but burns and melts in the flame. That melt accelerates fire spread.

Drawbacks — Tar and volatile oil products dissolve XPS. Rodents and insects devour it. But it is an effective electrical insulator.

Usage

Never use XPS in applications where temperatures above 70 °C are expected. The material is well suited as protected membrane roof insulation, cavity insulation, floor insulation, pitched roof insulation and perimeter insulation. It also figures as an excellent choice for inside insulation, on condition the internal lining is fire resisting.

Comment

The propellant Freon 12, a CFC used in the past, was very stable, chemically inert and had a very low thermal conductivity; but an ODP-value of 1 (ODP stands for Ozone Depleting Potential). HCFC's, with and ODP of 0.02 to 0.15 replaced it. Alternatives with zero ODP and very low GWP (GWP stands for Global Warming Potential), such as pentane, are expected to see their usage increasing.

Polyurethane- and polyisocyanurate (PUR and PIR)

Both are the only insulation materials produced chemically by isocyanate reacting with polyolefin in the presence of a catalyst, a propellant and the necessary additives. The difference between the two relates to the isocyanate ratio, in PIR high enough (60 to 65% kg/kg instead of 50 to 55% kg/kg) to form auto-polymers. As the explosive isocyanate/polyolefin reaction shows a high sensitivity to temperature and relative humidity, a strict control of both parameters is necessary. The reaction product is very sticky, which allows spraying the mixture on any kind of facings. Producing sandwich panels is no problem that way. Once the reaction is finished, boards are cut into the right size and stored. Until the early nineties Freon 11, with an ODP of 1, was used as propellant. Since it was replaced by lower ODP propellants.

Properties

Density — Must pass 30–32 kg/m³. If not, the foam remains quite unstable. An upper boundary is difficult to fix. Structural applications demand densities up to 60 kg/m³.

Thermal

Specific heat capacity — ≈ 1470 J/(kg·K)

Apparent thermal conductivity — With density at 10 °C: $-0.112 + 1.86 \cdot 10^{-3} \rho + 2.362/\rho$. The low value results from the propellant. Effusion induces a slow increase with age: from ± 0.018 W/(m·K) after production to an equilibrium value 0.023 W/(m·K). Aging speed depends among others on the diffusion resistance of the linings used.

Hygric

Moisture content — As continuous foams, PUR and PIR are hardly hygroscopic and non-capillary. Water heads give some moisture uptake depending on the percentage of open pores.

Vapour resistance factor — Open pores limit the average dry value to $1.7 \exp(0.088 \rho)$ for a density (ρ) above 20 kg/m³. That value drops a little at higher relative humidity.

Air — Both PUR and PIR are quite airtight.

Strength and stiffness — Compression resistance increases with density. 30 kg/m³ gives $\sigma_{10} \approx 0.15$–0.25 MPa at 20 °C. That value drops somewhat at higher temperatures. The allowable stress is $\sigma_{10}/3$.

2.3 Thermal insulation materials

Behaviour

Under mechanical load	Both show higher creep sensitivity with temperature.
Sensitivity to temperature, IR and UV	PUR and PIR are thermo hardening. They carbonize at high temperature. Both have a high thermal expansion coefficient, 45–$70 \cdot 10^{-6}$ K^{-1}. With too light foam, irreversible deformation may happen. PUR loses structural stability above 100 °C, PIR above 130 °C.
Under moisture load	Both PUR and PIR show good moisture tolerance though moisture in the pores in combination with high temperature fluctuations may increase dimensional instability, sometimes with annoying consequences as Figure 2.7 illustrates.
Exposure to fire	PUR shows worse fire response than PIR. Igniting PUR is difficult but once it burns, it sustains fire spread while carbonizing with emission of a black toxic smoke. PIR does better thanks to the addition of halogenated polyoles. It does not contribute to fire spread and carbonizes with little smoke.
Drawbacks	Rodents and insects devour both. As all plastic foams do, the two act as excellent electrical insulators.

Figure 2.7. Irreversible deformation.

Usage

PUR and PIR are fit for most applications. On site spraying is possible, allowing insulating the most complex shapes. Widely spread is on site sprayed floor insulation.

Pressed perlite boards (PPB)

Basic materials are cellulose fibres and perlite. The two are wet-mixed while adding a synthetic resin. Pressing, cooling and drying gives dense boards, available in two thicknesses: 20 and 40 mm. Greater thicknesses are realized gluing boards on top of each other.

Properties

Density	Between 130 and 215 kg/m^3. This is heavy for an insulation material.
Thermal	
Specific heat capacity	$\approx$ 1000 J/(kg·K)
Apparent thermal conductivity	With density at 20 °C: $0.0455 + 1.35 \cdot 10^{-4} (\rho - 100)$ The rather high value increases quite fast with volumetric moisture ratio, for boards of 142 kg/m^3 at 20 °C: $0.0525 + 0.00177\, \Psi$ (Ψ in % m^3/m^3)
Hygric	
Moisture content	PPB is hygroscopic, sorption being given by (ϕ in %, Ψ in % m^3/m^3): $\Psi = 0.1\, \phi\, / (-0.00193\, \phi^2 + 0.2052\, \phi + 2.793)$. The boards show some capillarity and get wet under water head. Long lasting submersion soaks them (till 55 % m^3/m^3).
Vapour resistance factor	Open porosity results in low values: $1 / [0.13 + 0.193\, (\phi/100)^4]$, ϕ being relative humidity in %
Air	PPB is air permeable.
Strength and stiffness	Shows decent compression resistance but hardly tensile strength.

Behaviour

Sensitivity to temperature, IR and UV	The boards are extremely temperature tolerant.
Under moisture load	The synthetic resin used hydrolyses when wet. This further degrades the already low tensile strength.
Exposure to fire	PPB is non-combustible.

Usage

The material was used as low-sloped roof insulation. Actual insulation requirements, however, are so strict that PPB is hardly used as an insulation material anymore.

2.3.4.3 Insulating systems

Radiant barriers are an example. Typical layout is an air bubble foil, covered at both sides with reflective linings. When such foils face an air cavity, cavity thermal resistance increases though final value will depend on mean temperature, temperature difference between surfaces, cavity slope, heat flow direction and overall air-tightness of the radiant barrier system. As an example, Figure 2.8 shows the thermal resistance of a horizontal cavity with a radiant barrier, long wave emissivity 0.1, at one side, the heat flowing bottom-up, the cavity situated behind the outer finish, inside temperature 20 °C and the inside leaf having a thermal resistance of 4 m^2·K/W. The increase in cavity thermal resistance with outside temperature seems not negligible. Of course, compared to the inside leaf's 4 m^2·K/W, it is of no importance.

With that one example, it must be clear that manufacturers who claim constant thermal resistance upgrades when using their reflective systems comparable with applying 20 cm of mineral wool are misleading customers.

2.3 Thermal insulation materials

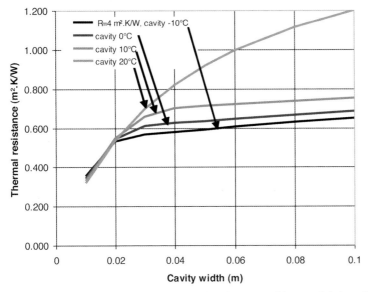

Figure 2.8. Horizontal air cavity, radiant barrier at one side, $e_L = 0.1$, heat flow bottom-up, cavity behind outer finish, temperature indoors 20 °C, temperature outdoors –10, 0, 10 and 20 °C, thermal resistance of the inside leaf 4 m² · K/W.

2.3.4.4 Recent developments

The insulation materials discussed above were all developed before 1960. Since then, manufacturers concentrate on better properties and new applications. Their apparent thermal conductivity anyhow has one characteristic in common: it becomes significant second place after the decimal point. Since 1960, there have been two new developments: transparent and vacuum insulation.

Transparent insulation (TIM)

A typical transparent insulation panel consists of synthetic colourless straws bundled in parallel, covered on both sides by a glass sheet and the perimeter sealed. The material couples transparency for short wave radiation with low heat conductivity, hardly any convection in the straws and little transmission of long wave radiation. To get optimal profit of transmitted short wave radiation, the surface behind the panels must be black-coloured. The advantage of TIM should be a better overall annual heat balance than with classic insulation materials.

Properties

Thermal

Apparent thermal conductivity   Differs between orthogonal and parallel to the synthetic transparent straw bundle
Orthogonal: $0.055 - 0.073\,d + 3.4 \cdot 10^{-4}\,\theta$
Parallel: $0.061 + 0.285\,d + 7.1 \cdot 10^{-4}\,\theta$

Short wave transmissivity   Changes with solar incidence (χ):
$0.94 - 3.8 \cdot 10^{-4}\,\chi - 6.95 \cdot 10^{-5}\,\chi^2$

Hygric

Moisture content	–
Vapour resistance	The glass sheets and perimeter seal should give unlimited vapour resistance.
Air	TIM panels should be airtight.

Behaviour

Sensitivity to temperature, IR and UV	The synthetic transparent straw bundle suffers from discoloration under UV-radiation.
Under moisture load	May give problems if the perimeter seal leaks. Then water vapour will diffuse into it and dust will enter the panels, resulting in condensation and dust deposit against the backside of the coldest glass sheet.
Exposure to fire	The synthetic straw bundle melts.
Drawbacks	Overheating. Solutions proposed are inclusion of solar shading in the TIM-panels, adding solar shading to the TIM-panels or leaving a ventilated air cavity behind, drawing the heated ventilation air into the building when advantageous but venting it to the outside when needed to avoid overheating.

Usage

TIM did not see widespread application. One reason is cost. Necessary solar shading in fact increases the TIM investment to a level that it stops being competitive with classic insulation materials. The few applications also had problems with yellowing of the panels

Vacuum insulation (VIP)

VIP typically is manufactured by enveloping micro porous fumed silica boards with a gas-tight facing, followed by vacuum evacuation of the fumed silica. That way conduction in the pores is largely eliminated, turning conduction along the pore walls and long wave radiation in the pores into the only heat transfer modes left. Facings used are multilayer metalized polymer films or thin metal films.

Properties

Thermal

Apparent thermal conductivity	Distinction must be made between the central apparent and overall equivalent thermal conductivity. For new VIP's, the first is as low as 0.003 W/(m · K), but slow aging caused by air permeating across the facing into the pores raises that value to 0.006–0.01 W/(m · K). Facings also act as thermal bridges along the panel's perimeter with the linear thermal transmittance for aluminium film as shown in Figure 2.9.

Hygric

Vapour resistance	VIP's normally have an unlimited vapour resistance.
Air	VIP's should be airtight.

2.4 Water, vapour and air flow control layers

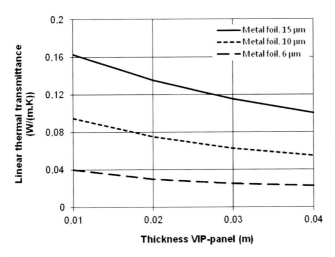

Figure 2.9. VIP-panel with thin aluminium film facing: perimeter thermal bridge.

Behaviour

Sensitivity to temperature, IR and UV
As long wave radiation is the main heat transfer mode defining apparent thermal conductivity, the value increases with temperature.

Usage

Applying VIP's in buildings is not straightforward. Perforation of the facing must be avoided. Cutting panels into the right format is excluded. Use in prefabricated panels could be a possibility. Some also see low-sloped roof, terrace and floor insulation as an alternative, on the condition the boards are protected by additional 'classic' insulation. Each application anyhow requires end and corner panels with deviating dimensions.

Comment

An alternative for VIP's are plastic foams with nanopores. In fact, in pores with dimensions close to the free path length of the pore gas molecules, conduction in the gas goes to zero, leaving matrix conduction and long wave radiation as the main heat transfer modes. Radiation can be minimized by adding graphite. As for VIP, apparent thermal conductivity may drop as low as 0.004 W/(m·K), while cutting and perforating should not create problems.

2.4 Water, vapour and air flow control layers

2.4.1 In general

While apparent thermal conductivity directs developments in insulating materials, water permeability, vapour permeability and air permeability do it for moisture, water vapour and airflow control layers, commonly called retarders and barriers. The three properties should have values close to zero. Permeability depends on the material's open porosity and the equivalent pore diameters. An open porosity of zero allows neither flow nor diffusion. With open pores so small that fluids and gasses experience extreme friction when passing, flow will be negligible. In truly very small pores, Fickian diffusion even turns into Knudsen diffusion:

$$G_v = -0.41\, d^3 \sqrt{\frac{1}{R\,T}}\, \text{grad}(p_v) \tag{2.14}$$

where d is the equivalent pore diameter. Knudsen diffusion gives very low water vapour permeability (for $d = 10^{-9}$ m a value $1.4 \cdot 10^{-12}$ s). Barrier layers therefore have an open porosity of zero or pores so small they retard water flow, airflow and vapour diffusion to the maximum. Absolute barriers have air and water permeability of zero along with an infinite diffusion resistance.

Other properties and overall behaviour of course are also important, like thickness, strength, chemical resistance, etc. In addition, ease of installation plays a role.

2.4.2 Water barriers

2.4.2.1 A short history

Until the 1930's, waterproofing of structures was done with tar felt drenched with volcanic cement. Tar, a mixture of hydrocarbons, was a by-product of the coke production. It showed high resistance against root perforation, had self-curing capabilities and was cheap. Temperature and UV tolerance however was less. After World War II, bitumen replaced tar. Soft bitumen is the heaviest cracked crude fraction, a mixture of light and heavy hydrocarbons. Oxidation allows increasing the molecular weight of the lightest components, which results in oxidized bitumen. The advantages of oxidized bitumen compared to tar are a higher hardness and softening point, more temperature and UV tolerance and less flow.

The first bituminous product marketed was naked bituminous felt, an organic felt drenched in soft bitumen. The product was initially used as an insert for bitumen poured or brushed on site. The idea quickly occurred to cloak the naked felt with oxidized bitumen during manufacturing and to sell it by roll. Coated bituminous membrane, called 'roll-roofing' was born. Application consisted of adhering a first layer with hot bitumen on the substrate, followed by two additional layers adhered on top of the other. The next step was the introduction of so called 'burn rolls' or millimetre roofing. The bitumen for adhering coated membranes was added in the factory. Application was now possible using the gas flame. In cases higher tensile strength was needed, jute replaced the felt insert.

But bitumen with felt or jute inserts caused problems. Both inserts pick up moisture. As a result, the built-up roofing developed micro-bubbles that crack when stepped on. Organic inserts could also rot whereas their deformability together with the elastic/plastic response of oxidized bitumen promoted blistering. When low-slope roofs started to be insulated, increased temperature load on the roofing saw these disadvantages multiply, which is why beginning in the sixties felt and jute were replaced by glass fibre and glass fabric. That generated new drawbacks. The use of these elastic materials without plasticity increased ripping sensitivity of the roofing. For that reason, polyester fibre felt and fabric started replacing glass fibre felt and fabric in the eighties.

2.4 Water, vapour and air flow control layers

Meanwhile, research showed that aging oxidized bitumen was at a disadvantage. Temperature and UV tolerance is too low to function properly for a long enough period as roofing membrane on insulated low-slope roofs, a fact accelerating the introduction of new membranes: polymer-bitumen with polyester insert and polymers.

2.4.2.2 Bituminous membranes

All bituminous membranes consist of an insert, at both sides enrobed with oxidized bitumen. A classification with the characteristics listed in Table 2.2 is:

Bitumen	Insert	
	Glass fibre felt/fabric	**Polyester fibre felt/fabric**
Coated	With glass fibre felt/fabric VP50/16	With polyester fibre felt/fabric P150/16
	Idem gravelled top layer VD45/30	
	Idem, perforated VP45/30	
	Idem, perforated VP40/15	
Burn-roll	With glass fibre felt/fabric V3	With polyester fibre felt/fabric P3
	Idem, V4 (3 and 4: thickness in mm)	Idem, P4 (3 and 4: thickness in mm)

The mechanical properties of bitumen are temperature related. When warm, bitumen behaves as a viscous liquid; while cold, it turns hard and brittle. Aging under temperature, humidity and UV load occurs quite fast, first by loss in flexibility, later followed by cracking. For these reasons, bituminous membranes should only be used as base or intermediate layer in built-up roofing on insulated surfaces. Perforated glass fibre bitumen VP45/30 is used as base layer when partial adherence using hot bitumen is preferred while perforated glass fibre bitumen VP40/15 is applied as a base layer for partial adherence using the gas flame. Adhering glass fibre felt burn-roll bitumen might be done using hot bitumen or the gas flame. Gluing with polymeric resin and self-adhering bitumen were recently introduced. Polyester burn-roll bitumen P150/16, P3 or P4 deserves recommendation when extra punch resistance is needed.

2.4.2.3 Polymer-bituminous membranes

Polymer-bitumen membranes consist of an oxidized bitumen/polymers mixture cloaking a polyester fibre felt/fabric insert. Thickness is 3 or 4 mm. Two types of polymers are commonly used: SBS and APP. SBS polymer-bitumen contains some 12% kg/kg Styrene/Butadiene/Styrene (SBS) elastomeric and some 88% kg/kg oxidized bitumen. The SBS elastomeric embeds a three dimensional grid in the bitumen thereby improving elasticity. The mechanical properties are more temperature tolerant and aging under heat and moisture load develops much slower than with oxidized bitumen. Only UV tolerance poses problems. A surface protection with gravel or slate shipping therefore is necessary. APP polymer-bitumen contains some 30% kg/kg Ataxic Polypropylene (APP) plastomeric and 70% kg/kg oxidized bitumen, with APP functioning as fibrous reinforcement. That way, APP shows the same plastic/elastic behaviour as oxidized bitumen. However, mechanical properties show more temperature tolerance and aging under heat, moisture and UV slows down.

For some of the properties, see Table 2.3. Adhering SBS and APP polymer-bitumen does not differ from adhering oxidized bitumen (hot bitumen, gas flame, gluing, self-adhering).

Table 2.2. Characteristics of bituminous membranes.

Characteristics	VP 50/16	VP 45/30	VP 40/15	V3	V4	P 150/16	EP2	P3	P4
Surface weight, g/m²	45–50	30–45	30–40	45–50	45–50	135–150	135–150	135–150	135–150
Perforations, diameter (mm)		19	40						
Perforated surface, % per m²		3–6	12–18						
Composition									
Bitumen amount, g/m²	750	700	700	2100	2700	750	1100	2100	2700
Filler content, % kg/kg	30	30	30	30	30	30	30	30	30
Min. weight, g/m²	1600	3000	1500	3000	4000	1600	1250	3000	4000
Top finish: sand/talc	x	x	x	x	x	x	x	x	x
burn off foil		x	x	x	x	x	x	x	X
Underside finish: sand/talc	x		x	x	x	x	x	x	x
burn off foil		x	x	x	x	x	x	x	X
gravel 2/4, 4/6		x							
Sand 1/3		x	x						
Thermal properties									
Dripping temp.[1]	70			70	70	70	70	70	70
Bending temp.[2] Cracking, θ (°C) <	20			20	20	20	20	20	20
Breaking, θ (°C) <	3			3	3	3	3	3	3
Mechanical properties									
Shrinkage (%)						≤ 0.5	≤ 0.5	≤ 0.5	≤ 0.5
Punch resist. (N)						≥ 100	≥ 100	≥ 100	≥ 100
Tensile strength, N per 50 mm	120–250	120–250	120–250	120–250	120–250	500–1200	500–1200	500–1200	500–1200
Application									
Base layer		x	x	x	x	x	x	x	x
Vapour barrier	x			x	x	x	x	x	x

[1] lower
[2] higher threshold
N Newton

2.4 Water, vapour and air flow control layers

Table 2.3. APP- and SBS-polymer-bituminous membranes: properties and behaviour.

Property	Value/behaviour
Tensile strength	500 to 1200 N per 50 mm
Strain at rupture	> 20%
Dripping temperature	110 to 150 °C
Bending test, cracking (temperature lower than)	−10 tot −25 °C
Aging	Slow, polymer-bitumen membranes perform well for a long period of time

2.4.2.4 High-polymer membranes

High-polymer membranes are again more temperature tolerant than oxidized bitumen. They also age slower. But, unlike polymer-bitumen, they require proper adhering techniques, demanding a specialized workforce. All are by-products of the petro-industry. Three groups can be distinguishes: elastomers, thermoplastic elastomers and plastomers. Elastomers consist of vulcanised hydrocarbons, plastomers of non-vulcanized hydrocarbons.

Most applied plastomer membranes		
PIB	Polyisobuthylene	Moderately weather resistant Sensitive to organic solvents Punch resistance upgraded by an insert at the backside Adhered with hot bitumen
PVC	Polyvinylchloride	If non-stabilized very UV sensitive Sensitive to bitumen and organic solvents Punch resistance upgraded by an insert at the backside Loosely laid, overlaps sealed with hot air or swell welding
PVF	Polyvinylfluoride	Highly UV tolerant No sensitivity to bitumen and organic solvents Glued. Overlaps sealed with self-adhering tape
Most applied thermoplastic elastomer membranes		
TPV	Thermoplastic vulcanised elastomer	Available as 1.2 mm thick membranes with or without insert
TPO	Thermoplastic polyolefine	Available as 1.2 mm thick membranes with or without insert
Most applied elastomer membranes		
IIR	Butyl rubber	Sensitive to organic solvents Extremely high water vapour resistance Loosely laid or fixed by melting or with contact glue, overlaps glued
EPDM	Ethylene propylene copolymer and diene monomer	Sensitive to organic solvents Less vapour-tight than IIR Loosely laid or fixed by melting or with contact glue, overlaps glued

Thermoplastic elastomers sit in between. Elastomers react elastically up to about 300 °C. Thermoplastic elastomers behave elastically up to about 200 °C and show plasticity above. Plastomers behave plastically from 120 °C on, which eases adaptation to any kind of substrate relief. Plasticity however also means the membrane may suffer from lasting deformations and low punch resistance. For that, plastomers are equipped with a glass and/or polyester fabric insert. Overall UV-sensitivity of high-polymer membranes is moderated by adding stabilisers.

2.4.3 Vapour retarders and vapour barriers

Vapour retarders and vapour barriers have a high vapour resistance. In theory, this could be achieved by using thick layers of a material with moderate or even quite low vapour resistance factor. Such a solution however is neither buildable nor affordable. Thin foils with very high vapour resistance factor are a logical choice instead. A high vapour resistance factor means an open porosity near zero. Some paints meet that requirement, as do synthetic foils, bituminous membranes and metallic foils. The sequence is: $[\mu d]_{o,paint} < [\mu d]_{o,synthetic} < [\mu d]_{o,bitumen} < [\mu d]_{o,metal}$ with d thickness and μ vapour resistance factor.

To bring some order in the overall offer, vapour retarder and barrier classes have been proposed with the equivalent diffusion thickness as a variable $[\mu d]_{eq}$, 'equivalent' because real diffusion thickness largely depends on workmanship. Take for example a foil with thickness d and vapour resistance factor μ_o. Assume poor workmanship resulting in $p\%$ perforations. Equivalent diffusion thickness $[\mu d]_{eq}$ than becomes:

$$[\mu d]_{eq} = \frac{1}{(1-p/100)/[\mu d]_o + p/(100\,d)} \quad (2.17)$$

Figure 2.10 translates this equation in a graph. The impact of perforations looks dramatic, especially for very vapour tight foils. When diffusion thickness is infinite, the equivalent value drops to $100\,d/p$, meaning that 1% perforations reduces equivalent diffusion thickness to 1.5 cm for a 0.15 mm thick foil!

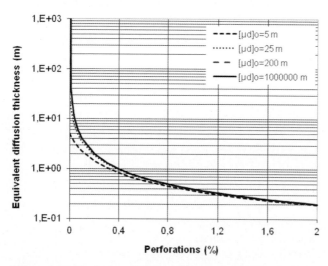

Figure 2.10. Equivalent diffusion thickness depending on percentage of perforations.

2.4 Water, vapour and air flow control layers

In Belgium, classification looks as follows:

Class	Boundaries	Examples
E1	$2\ m \leq [\mu d]_{eq} < 5\ m$	Bituminous craft paper with overlapping flanges taped Gypsum board with aluminium foil finished backside
E2	$5\ m \leq [\mu d]_{eq} < 25\ m$	Synthetic foils, $d = 0.2$ mm, mounted with taped overlaps
E3	$25\ m \leq [\mu d]_{eq} < 200\ m$	Bituminous, polymer bituminous, and high-polymer membranes, continuous polyester laminates
E4	$[\mu d]_{eq} \geq 200\ m$	Metallic foils coated with bitumen, oxidized bituminous membrane with 100 µm thick aluminium insert

Vapour retarders of class E1 and E2 can be stapled against studs, purlins and others. Instead, vapour barriers of class E3 and E4 must be adhered leak free to a substrate the same way as is done with roll-roofing. The class required for a given application figures as a performance requirement and should be listed in the specifications. In North America, three classes are defined by the international code:

Class	Boundaries
I	$[\mu d]_{eq} > 32\ m$
II	$3.2\ m \leq [\mu d]_{eq} < 32\ m$
III	$0.32\ m \leq [\mu d]_{eq} < 3.2\ m$

In the last decades, smart vapour retarders entered the market. Some have a diffusion thickness decreasing with relative humidity, an example being polyamide foil (Figure 2.11). Others impede vapour transfer but conduct water condensing on the retarder to the other side where it evaporates.

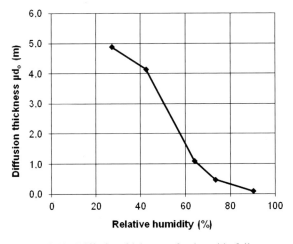

Figure 2.11. Diffusion thickness of polyamide foil.

2.4.4 Air barriers

Air barriers have very low air permeance. This could be realized using thick layers with moderate to quite high air permeability, but again logic prefers thin foils with very low air permeability. Low air permeability demands materials with no porosity, with closed pores or with a network of very fine pores. Those conditions are not as stringent as for a high vapour resistance factor, which demands open porosity near zero.

Nevertheless, the same materials are used: vapour barring paints, synthetic foils, bituminous membranes and metallic foils. On condition workmanship is perfect and no leaks are left, they provide vapour and air tightness, which is why the expression 'air and vapour retarder' is commonly used. In cases no vapour retarder is needed, a continuous layer of structured paint or a continuous plaster finish indoors may assure air-tightness. The details are of course critical: window reveals, skirting boards, electricity sockets, the wall/roof interfaces, etc. In fact, air barriers should be continuous. When airtight, tensile strength, ripping strength or adhering strength must also be sufficient to withstand wind pressure and stack effect. If not, a continuous support has to be provided. Other properties concern durability and fire resistance.

In Canada, a difference is made between four classes of air barriers:

Class	Air leakage at 75 Pa $l/(s \cdot m^2)$
1	0.05
2	0.10
3	0.15
4	0.20

What class to choose depends on the equivalent diffusion thickness of the outer layer of the assembly. Class 4 applies if that equivalent diffusion thickness is lower than 0.3 m, class 1 if it is higher than 3 m. The two intermediate classes fit for equivalent diffusion thickness steps in-between: class 3 for 0.3 to 1.5 m and class 2 for 1.5 to 3 m. The acceptable air leakage of an air barrier material is set equal to 0.02 $l/(s \cdot m^2)$ at an air pressure difference of 75 Pa.

2.5 Joints

2.5.1 In general

Joints are a necessity in building constructions. Mounting prefabricated panels for example demands larger openings than the dimensions of the panel, included the tolerances. Joints compensate necessary differences in dimensions. In fact, if a window frame should have identical exterior dimensions as the bay it has to fill, mounting becomes impossible. Height and width of the frame should thus be:

$$[h_{windoe}, b_{window}] = [h_{bay} - \partial h_{bay}, b_{bay} - \partial b_{bay}] \tag{2.16}$$

with ∂h_{bay} and ∂b_{bay} joint width between window and bay.

2.5 Joints

Buildings composed of volumes with different heights or bearing varying loads and buildings constructed on soils with variable compressibility may suffer from differential settling. To avoid cracking, foundation to roof settlement joints are needed. Building part dimensions also change with temperature and relative humidity. To avoid random cracking, expansion joints at well-planned locations are necessary. The same holds for cement-based materials suffering from shrinkage during binding. Also there, well-positioned joints must avoid random cracking. In both cases, random cracks in envelopes may disturb rain proofing and air tightness.

2.5.2 Joint solutions and joint finishing options

Common joint solutions are:

One step	The outermost sealant cares for wind and rain tightness (Figure 2.12).
Two steps	The joint gets a wind tight sealant indoors and a rain tight profiling outdoors with a pressure equalizing chamber in between (Figure 2.13).

As joint finishing options, one has:

Sealants	They consist of an organic binder, inert filler and additives. Together these shape the properties. Non-setting sealants are gun mouldable. Afterwards they solidify or stay mouldable. Elastomeric ones are mouldable when gunned but polymerise after.
Preformed profiles	Are designed to stretch when pressed in the joint. That compresses them, closing the joint (Figure 2.14).

Let's take a look now at sealants.

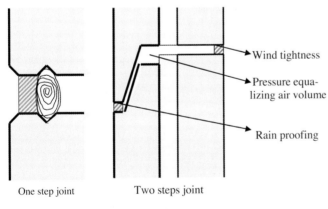

Figure 2.12. One step joint

Figure 2.13. Two steps joint

Figure 2.14. Pre-formed profile.

2.5.3 Performance requirements

2.5.3.1 Mechanical

Also when aged, sealants must still compensate movement, remain adhering, stay rain proof and be wind tight, whereas damage tolerance should remain high. These performances translate into the following requirements:

1. Sufficient adherence — As the upper boundary of adherence is tensile strength, the requirement is 'sealants must keep sufficient tensile strength'.
2. Good deformability — Ideal sealants should absorb any joint edge movement without coherence forces. Deformability is judged by testing resilient recovery after straining and measuring fracture strain after aging.
3. Correct hardness — Too soft sealants are very damage sensitive. Hardness is judged by measuring the modulus of elasticity.

2.5.3.2 Building physics related

Two requirements surface: (1) a joint must stay rain proof and wind tight during service life, (2) unacceptable interstitial condensation at the back of the front seal must be excluded. For a two step solution, a direct consequence in moderate and cold climates of this second requirement is the inner wind tight seal must have a higher diffusion resistance than the outer rain proofing. However, at the back of one-step joints, interstitial condensation is unavoidable. The measure needed there consists of designing a problem-free condensate discharge.

2.5.4 Sealant classification

ISO 11600 differentiates between glass (G) and building (F) sealants. For both, the strain in percentage absorbed after artificial aging figures as a classification pivot:

	Glass sealants (G)		Building sealants (F)		
Elastomeric sealants	Class 25	25 LM	Class 25	25 LM	
		25 HM		25 HM	
	Class 20	20 LM	Class 20	20 LM	
		20 HM		20 HM	
			Class 12.5	12.5 LM	
				12.5 HM	Non setting sealants
			Class 7.5		

Class 25 means 25% deformability whereas the addition LM of HM stands for high or low modulus of elasticity. While elastomeric sealants show high deformability, for non-setting sealants deformability is less but hardness scores higher. For a sealant to belong to one of the classes, the requirements listed in Tables 2.4 and 2.5 must be fulfilled (testing according to ISO standards).

2.5 Joints

Table 2.4. Glass sealants (G), properties.

Property	Class			
	25 LM	25 HM	20 LM	20 HM
Resilient recovery (% m/m)	≥ 60	≥ 60	≥ 60	≥ 60
Modulus of elasticity (N/mm²)				
• at 23 °C	≤ 0.4	> 0.4	≤ 0.4	> 0.4
• and/or at –20 °C	≤ 0.6	> 0.6	≤ 0.6	> 0.6
• for a strain of (% m/m)	200	200	160	160
Loss in volume (%)	≤ 10	≤ 10	≤ 10	≤ 10
Flow at 5 and 50 °C (mm)	< 3	< 3	< 3	< 3
Adhesion/cohesion at varying temperatures, with exposure to UV and plastically elongated after submersion in water	Deficiencies over less than 5% of their length for 3 test samples. After a second test, none of the 3 samples may show deficiencies over more than 10% of their length			

Table 2.5. Building sealants (F), properties.

Property	Class						
	25 LM	25 HM	20 LM	20 HM	12.5 LM	12.5 HM	7.5
Resilient recovery (% m/m)	≥ 70	≥ 70	≥ 60	≥ 60	≥ 40	No requirements	
Modulus of elasticity (N/mm²)							
• at 23 °C	≤ 0.4	> 0.4	≤ 0.4	> 0.4	—		
• and/or at –20 °C	≤ 0.6	> 0.6	≤ 0.6	> 0.6	—		
• for a strain of (% m/m)	200	200	160	160	—		
Fracture strain	—	—	—	—	—	≥ 100	≥ 20
Adhesion/cohesion after submersion in water, fracture strain (%)	—	—	—	—	—	≥ 100	≥ 20
Loss in volume (%)	≤ 10	≤ 10	≤ 10	≤ 10	≤ 25	≤ 25	≤ 25
Flow at 5 and 50 °C (mm)	< 3	< 3	< 3	< 3	< 3	< 3	< 3
Adhesion/cohesion at varying temperatures, plastic elongation	Deficiencies over less than 5% of their length for 3 test samples. After a second test, none of the 3 samples may show deficiencies over more than 10% of their length						

2.5.5 Load and sealant choice

The remaining question is where to apply which sealant classes. The load, which depends on four parameters, gives the answer:

1. Outside environment (chemical aggressiveness)
2. Traffic at the building location (vibrations)
3. Height above grade (wind and rain)
4. Facade relief (wind and rain)

That results in the three load categories of Table 2.6. For the term 'receding', see Figure 2.15 (recess divided by height beyond 0.4). The table also gives the sealant classes per load for several applications. Non-setting sealants hardly show applicability. From load two on, a sealant class 20 or better must be used. In two-step joints, wind tightness is considered a load 1 application. Sealants of class 20 or better are therefore recommended.

Table 2.6. Loads and sealant choice per load.

Outside environment	Facade lay out	Traffic	Load, F (height above grade)		
			0–15 m	15–140 m	> 40 m
Non aggressive	Receding	Normal	1	1	2
		Busy	1	2	3
	Exposed	Normal	1	2	3
		Busy	2	3	3
Aggressive	Receding	Normal	1	2	3
		Busy	2	3	3
	Exposed	Normal	2	3	3
		Busy	3	3	3

Load	Glass sealants (G) $U_{glass} \leq 2.8$ W/(m² · K)						Building sealants (F)		
	Heavy facade walls		Panel facade walls		Lightweight facade walls		Prefabricated envelope parts		Settlement and expansion joints
	CG[1]	CoG[2]	CG	CoG	CG	CoG	$h <$ story	$h >$ story	Curtain walls
1	20 HM	20 LM	20 LM	25 HM	25 HM	25 HM	12.5 LM	25 HM	25 LM
2	25 HM	25 HM	25 LM	25 LM	25 LM	25 LM	25 HM	25 LM	25 LM
3	25 HM	25 LM	25 LM	25 LM	25 LM	25 LM	25 HM	25 LM	25 LM

[1] CG: clear glass
[2] CoG: coloured glass

2.5 Joints

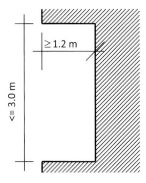

Figure 2.15. Receded facade.

2.5.6 Structural design of sealed joints

Assembling joints, settlement joints and expansion joints have to withstand tension, compression and shear forces. For the maximum allowed strain under tension and compression, see Table 2.7 and Figure 2.16. These values have to be multiplied with 1.5 in case of shear.

Deformation is a consequence of hygrothermal movement, shrinkage and differential settlement of the joint edges. Assume tension or compression to be active. If $\Delta L_1 + \Delta L_2$ is the maximum displacement of the joint edges compared to their initial position, then joint width should equal:

$$d_{\text{joint}} = 100\left(\Delta L_1 + \Delta L_2\right)/\varepsilon \tag{2.17}$$

Table 2.7. Tension and pressure, allowable strain.

Class	Allowed strain ε (%)
7.5	≤ 7.5
12.5 HM	≤ 12.5
12.5 LM	≤ 12.5
20 HM	≤ 20
20 LM	≤ 20
25 HM	≤ 25
25 LM	≤ 25

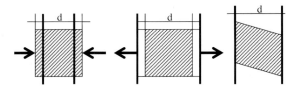

Figure 2.16. Sealants: strain linked to pressure, tension and shear loads.

If only shear is active and gives a maximum parallel displacement as compared to the initial position, then joint width should be:

$$d_{joint} = 10\,\Delta s / (3\,\varepsilon)^{0.5} \qquad (2.18)$$

If the three act together ($\Delta L_1 + \Delta L_2 = \Delta L$), joint width should become:

$$d_{joint} = \max\left[100\,\Delta L/\varepsilon,\; \frac{16.7\,\Delta L}{\varepsilon}\left(\frac{133}{\varepsilon} + \sqrt{1+\left(\frac{\Delta s}{\Delta L}\right)^2}\right)\right] \qquad (2.19)$$

The three formulas do not account for tolerance. For a joint, whose designed width is b, real width will lay between $b - (t_1 + t_2)$ and $b + (t_1 + t_2)$ with t_1 and t_2 the 95% limits of expected dimensional tolerance for both parts forming the joint. If in reality the width is smaller than designed, then the sealant will show higher deformation with adhesion losses or splitting as attendant problems. That requires the 95% limit of the expected tolerances to be added to the calculated joint width. Width of non-setting sealants is best limited to 20 mm. Elastomeric sealants can tighten any joint width, though widths beyond 30 to 40 mm are not recommended.

2.5.7 Points of attention

Sealants demand dry weather and outdoor temperatures above 2 °C when applied. Freezing, wet weather or mist requires specific precautions such as working under a tent. First, a foam rubber or self-swelling strip is pressed in the joint, a job facilitated if grooves are present at some distance from the outer surface. Then the edges are cleaned of dust and fat. If needed, they get a primer or pore filler treatment as recommended by the sealant manufacturer. After, the sealant is spouted and smoothed. For the sealant's thickness, the following rules apply:

- Non setting sealants 7.5 and 12.5 HM:
  Thickness and width equal
- Elastomeric sealants 12.5 LM, 20 HM, 20 LM, 25 HM, 25 LM:
  Thickness half the width

The surfaces elastomeric sealants adhere to must be parallel. That way, stress and strain remains linear. Filling corners, adhering to oblique surfaces or tightening smaller joints are faulty applications (Figure 2.17). One should also look to incompatibilities. Sealants may react with the materials they adhere too. Manufacturers should give the necessary information.

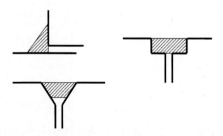

Figure 2.17. Elastomeric sealants faulty applied.

2.6 References and literature

[2.1] Griffin, C. W. (1970). *Manual of build up roof systems.* Mac Graw-Hill Book Company.

[2.2] Mathey, R., Cullen, W. (1974). *Preliminary Performance Criteria for Bituminous Membrane Roofing.* Paper 48 of the International Symposium on Roofs and Roofing, Brighton.

[2.3] Isaksen & Juul (1974). *Bitumen felt roofings on polystyrene insulated roofs.* Paper 2.5.1 of the Second CIB-RILEM Symposium on Moisture Problems, in Buildings, Rotterdam.

[2.4] WTCB (1975). *Technische voorlichting 107.* Dichtingsmastieken voor gevels, 22 p. (in Dutch).

[2.5] WTCB (1976). *Technische voorlichting 110.* Dichtingsprofielen, 23 p. (in Dutch).

[2.6] WTCB (1978). *Technische voorlichting 124.* Kitvoegen tussen gevelelementen, 24 p. (in Dutch).

[2.7] Carpentier, G. (1976–1985). *De hygrothermische eigenschappen van bouw- en isolatiematerialen en het grensgedrag in aanwezigheid van vocht.* TI-KVIV, Cursus 'Thermische isolatie en vochtproblemen in gebouwen' (in Dutch).

[2.8] Bomberg, M., Shirtliffe, C. J. (1978). *Blown cellulose fiber thermal insulation.* ASTM Special Technical Publication No. 660.

[2.9] CRIR (1979). *Isolation thermique et humidité.* Rantigny (in French).

[2.10] Vos, B. H., Tammes, E. (1980). *Warmte- en Vochttransport in Bouwconstructies.* Samson Uitgeverij, Alphen aan de Rijn (in Dutch).

[2.11] Hens, H., Bouwfysica (1978–1981 *Warmte en Vocht, Teoretische grondslagen.* 1e en 2e uitgave, ACCO, Leuven (in Dutch).

[2.12] Bomberg, M. (1980). *Glassfiber as insulation and drainage layer on exterior of basement walls, Thermal Insulation Performance.* ASTM, Special Technical publication No. 718, p. 57–76.

[2.13] Bomberg, M. (1980). *Some performance aspects of glassfiber on the outside of basement walls, Thermal Insulation Performance.* ASTM, Special Technical publication No. 718, p. 77–91.

[2.14] Bomberg, M. (1980). *Problems in predicting the thermal properties of faced polyurethane foams, Thermal Insulation Performance.* ASTM, Special Technical publication No. 718, p. 412–428.

[2.15] Höfer, G. (1981). Extrudierter Polystyrol-Schaumstoff in Prüfung und Praxis. *Die Bautechnik 58,* Heft 2, pp. 37–46 (in German).

[2.16] Mewis, J., Hens, H., Standaert, P., Wouters, P., Meire, C. (1982). *Warmteisolatie: materiaalkundige, bouwfysische, bouwkundige en technico-economische aspecten.* Nationaal Programma RD-Energie, Spa conference (in Dutch).

[2.17] Castagnetta, V. (1983). *Recenti sviluppi dei materiali bituminosi normale e modificati nella impermeabilizzatione e protezione delle opere idrauliche.* 5th conference of the Association Internationale de l'étanchéité, Strasbourg (in Italian).

[2.18] WTCB (1984). *Technische Voorlichting 151.* Soepele dakafdichtingsmaterialen, 24 pp. (in Dutch).

[2.19] BKV (1985). *Thermische isolatie: enkele praktische oplossingen aan de hand van kunststof-isolatie-materialen.* Brussel (in Dutch).

[2.20] University of Illinois (1986). *Small Homes Council-Building Research Council, Insulating Materials.* Council Notes Volume 8, Number 1, 8 pp.

[2.21] Teksten VDAB-cursus 'Isolatietechnieken' (1986–1991) (in Dutch).

[2.22] Fechiplast (1986). *Kunststoffen vandaag en morgen.* Brussel (in Dutch).

[2.23] Hens, H. (1987). *Buitenwandoplossingen voor de residentiële bouw: platte daken.* Eindrapport Nationaal Programma RD-Energie, 150 p. (in Dutch).

[2.24] TU-Delft (1987). *Bouwfysisch Tabellarium* (in Dutch).

[2.25] Timusk, J., Tenende, L. M. (1988). Mechanism of drainage and capillary rise in glas fibre insulation. *Journal of thermal insulation,* Vol. 11, April, pp. 231–241.

[2.26] NRC-IRC (1989). *An Air Barrier for the Building Envelope.* Proceedings of the Building Science Insights 86, 24 p.

[2.27] IEA-Annex 14 (1992). *'Condensation and Energy'.* Final report, Volume 3, Catalogue of Material Properties, ACCO, Leuven.

[2.28] Proceedings of the Vth DOE-ASHRAE-BETEC Performances of the Exterior Envelopes of Buildings Conference. Clearwater Beach, December 1992.

[2.29] Sauerbrunn, I. (1993). *Dämmstoffe für den baulichen Wärmeschutz.* DBZ 12 (in German).

[2.30] Epple, H., Preisig, H. (1993). Zellulosefasern als Wärmedämmstoff in Hohlräumen. *Schweizer Ingenieur und Architekt,* Nr 32, p. 552–556 (in German).

[2.31] CMHC (1993). *Air Barrier Technology Update.*

[2.32] Carpentier, G. (1994). Kitten voor gebouwen, indeling en eisen volgens ISO 11600. *WTCB-Tijdschrift,* 2^e sem., pp. 40–41 (in Dutch).

[2.33] ORNL (1994). *A guidebook for Insulated Low-Slope Roof Systems.* Final report IEA Exco ECBCS Annex 19.

[2.34] Svennerstedt, S. (1995). *Settling of attic loose fill thermal insulation.* Lund University, Department of Building Technology, Report TVBH-3018, Lund.

[2.35] Svennerstedt, B. (1995). Analytical Models for Settling of Attic Loose-Fill Thermal Insulation. *Journal of Thermal Insulation and Building envelopes,* Vol. 19, October, pp. 189–201.

[2.36] Proceedings of the VIth DOE-ASHRAE-BETEC Performances of the Exterior Envelopes of Buildings Conference. Clearwater Beach, December 1995.

[2.37] Kumaran, K. (1996). *Final Report IEA-Annex 24, Material properties.* ACCO, Leuven.

[2.38] Laboratory of Building Physics, K. U. Leuven, Several Case Study and Forensic Investigation Reports, 1975–1996 (in Dutch).

[2.39] Künzel, H. (1997). Die Folie denkt mit. *Stuck-Putz-Trockenbau 55,* Heft 2, p. 34–37 (in German).

[2.40] Verbeeck, G., Hens, H., Constales, D., Van Keer, R. (1996). *Transparante isolatie: numerieke modellen en evaluatie van de prestatie van het system in het VLIET-HANTIE K30 proefgebouw.* REG-potentieel, 1e jaarverslag (in Dutch).

[2.41] Verbeeck, G., Hens, H., Constales, D., Van Keer, R. (1997). *Transparante isolatie: numerieke modellen en evaluatie van de prestatie van het system in het VLIET-HANTIE K30 proefgebouw.* REG-potentieel, 2e jaarverslag (in Dutch).

[2.42] Verbeeck, G., Hens, H., Constales, D., Van Keer, R. (1998). *Transparante isolatie: numerieke modellen en evaluatie van de prestatie van het system in het VLIET-HANTIE K30 proefgebouw.* REG-potentieel, 3e jaarverslag (in Dutch).

[2.43] Künzel, H. (1998). *The Smart Vapor retarder: An Innovation Inspired by Computer Simulations.* ASHRAE Transactions 1998, Volume 104, Part 2.

[2.44] Documentatiemappen van de producenten van isolatiematerialen, afdichtingsmaterialen en kitten (in Dutch).

[2.45] WTCB (2000). *Technische Voorlichting 213, Het platte dak.* Opbouw, materialen, uitvoering, onderhoud, 100 pp. (in Dutch).

[2.46] ASHRAE (2001). *Handbook of Fundamentals, Chapter 25.* Atlanta.

2.6 References and literature

[2.47] Künzel, H. (2001). *Trocknungsfördernde Dampfbremsen-Einsatzvoraussetzungen und feuchtetechnische Vorteile in der Praxis.* WKSB Heft 47, p. 15–23 (in German).

[2.48] Künzel, H., Leimer, H.-P. (2001). *Performance of Innovative Vapor Retarders Under Summer Conditions.* ASHRAE Transactions 2001, Volume 107, Part 1.

[2.49] Cauberg, J. J. M. (2002). Vacuümisolatie panelen, mogelijkheden voor de bouw. *Bouwfysica*, Vol. 12, nr 1, april, pp. 4–8 (in Dutch).

[2.50] Cauberg, J. J. M. (2003). Thermische kwaliteit van vacuümisolatie panelen. *Bouwfysica*, Vol. 13, nr 4, december, pp. 22–25 (in Dutch).

[2.51] Hens, H. (2004). *Toegepaste Bouwfysica 1, randvoorwaarden, prestaties, materiaaleigenschappen.* ACCO Leuven, 186 p. (in Dutch).

[2.52] Kwon, Y. C., Yarbrough, D. W. (2004). A Comparison of Korean Cellulose Insulation with Cellulose Insulation Manufactured in the United States of America. *Journal of Thermal Envelope & Building Science,* Vol. 27, No. 3, January.

[2.53] ASHRAE (2005). *Handbook of Fundamentals, Chapter 25.* Atlanta.

[2.54] Willems, W. M., Schild, K. (2006). *On the performance of vacuum insulated panels and sandwiches, Research in Building Physics and Building Engineering.* Taylor & Francis, London/Leiden/New York/Philadelphia/Singapore, 992 pp.

[2.55] ASHRAE (2009). *Handbook of Fundamentals, Chapter 26.* Atlanta.

[2.56] Baetens, R., Roels, S., Jelle, B., Gustavsen, A. (2010). *Long-therm thermal performance of vacuum insulation panels by dynamic climate simulations.* Proceedings of CESBP 2010.

[2.57] Hens, H. (2010). *Applied Building Physics, Boundary Conditions, Building Performance and Material Properties.* Ernst & Sohn (A Wiley Company), Berlin, 307 pp.

[2.58] Tenpierik, M., van der Spoel, W., Cauberg, H. (2010). Vacuümisolatiepanelen toegepast in de bouw, deel 1: thermisch gedrag. *Bouwfysica,* Vol. 21, nr 4, december, pp. 10–16 (in Dutch).

[2.59] Roels, S., Deurinck, M. (2011). The effect of a reflective underlay on the global thermal behaviour of pitched roofs. *Buildings and Environment,* Vol. 46, Issue 4, pp. 134–143.

3 Excavations and building pit

3.1 In general

With this chapter focus shifts to performance based design and construction of buildings. But before any building activity starts, one should test the soil's load bearing capacity using cone ground sounding. The results help in deciding how to excavate considering the number of below grade floors needed and how to lay the foundations. If the site is located between existing buildings, one needs a report describing their physical state before construction starts. That report will facilitate evaluation of possible future complaints by tenants of neighbouring premises about damage that could be the result of the building activity.

3.2 Realisation

Construction starts with demolishing existing buildings, if any, on the site, and organizing the cleared building site. In a next step, the contractor marks the building's footprint and fixes the reference level. Then excavating and stabilizing the building pit follows. Pits of limited depth in cohesive soils (clay, loam) can have vertical walls. A non-cohesive soil like sand demands sloped or buttressed walls, using for example 'Berliner wall' solutions (Figure 3.1). Buttressing is also needed when settlement around the pit must be excluded.

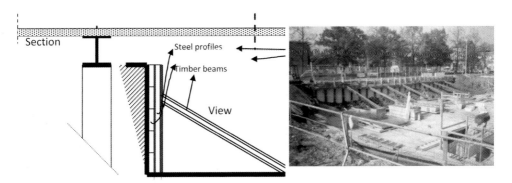

Figure 3.1. Buttressing using 'Berliner walls'.

In deeper building pits, problems with slide off and settlement are avoided by using sheet piles or deep walls. As excavation progresses, these are, if necessary, nailed in the ground beyond the farthest slide off plane using grout anchors (Figure 3.2). The number of anchors and anchor layers depends on pit depth, soil slide off sensitivity and surface loads outside the pit. In case excavation passes the water table, open drainage or well point systems are needed to lower it. In the last case, the radius of influence of the wells must be small enough to avoid differential settlement of nearby buildings. In order to ensure that does not happen, one can drive the sheet piles into the first impermeable soil layer below the water table or excavate the trenches for deep walls until such layer is reached.

Figure 3.2. Deep walls nailed with grout anchors.

When very deep basements in a densely built environment have to be excavated, an alternative solution after finishing the deep walls is to first cast the foundations in sheet-piled pits. Then one constructs all reinforced concrete columns from foundation to ground floor with all necessary floor dowels, followed by casting the ground floor, excavating top down each basement floor and casting each time the corresponding concrete floor. That way the successive floors serve as deep wall struts, making grout anchors redundant.

The excavated ground is transported to a land fill or bunkered on site for later usage. For more details, we refer to the specialized literature.

4 Foundations

4.1 In general

Foundations transfer the building load to the soil. Soil mechanics deliver the necessary knowledge to do that in a proper way. In this fourth chapter, the discussion is limited to a preliminary performance check, followed by a condensed overview of most common foundation systems. Also, some simple rules of thumb are given.

4.2 Performance evaluation

4.2.1 Structural integrity

In foundation design, two requirements prevail: (1) safety against soil fracture, (2) no unacceptable differential and overall settlement. The two translate into (1) the equilibrium and (2) the settling load bearing capacity of the soil.

4.2.1.1 Equilibrium load bearing capacity

Below any foundation, stress builds up three-dimensionally. The vertical component tends to move the soil upwards from below the footing. In case a fracture plane develops, sliding results. See Figure 4.1.

The load per square meter of contact area between footing and soil when that starts is called the equilibrium load bearing capacity, with a symbol p_s and units MPa. As Figure 4.1 shows, the

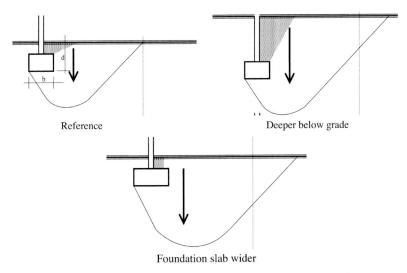

Figure 4.1. Equilibrium load-bearing capacity, sliding surface.

Performance Based Building Design 1. From Below Grade Construction to Cavity Walls.
First edition. Hugo Hens.
© 2012 Ernst & Sohn GmbH & Co. KG. Published 2012 by Ernst & Sohn GmbH & Co. KG

deeper below grade and the larger the contact area is, the higher the equilibrium load bearing capacity. In fact, in both cases, a larger ground mass opposes sliding. The value also increases with higher cohesion and grain compression in the soil. As a formula:

$$p_s = s_p \, \delta_p \, N_p \, \rho_k \, d + s_c \, \delta_c \, N_c \, c + s_\rho \, N_\rho \, \rho_k \, b \tag{4.1}$$

ρ_k is grain compression, c cohesion, d depth below grade and b the representative width of the footing. The constants N_p, N_c and N_ρ depend on the soil's internal friction angle (φ), see Table 4.1 and 4.2. The form factors s_p, s_c, s_ρ and depth factors δ_p, δ_c are given by:

$$\begin{aligned} s_p &= 1 + (b/L) \sin \varphi \\ s_c &= 1 + (b/L)(\pi + 2) \\ s_\rho &= 1 + 0.4 \, (b/L) \\ \delta_p &= 1 + \left[t \, g^2 \, (\pi/4 - \varphi/2) \exp(\pi \, t \, g \, \varphi) - 1 \right] \exp(-\pi \, R \, t \, g \, \varphi / d) \\ \delta_c &= \delta_p + (\delta_p - 1)/(N_p \, s_p - 1) \end{aligned} \tag{4.2}$$

with L the footing's length and R the hydraulic radius of the contact area with the soil. Allowable load bearing capacity is $p = p_s/n$, with n a safety factor, typically set equal to 2.

Table 4.1. Characteristics of some soils.

Soil	(Wet) density kg/m³	Cohesion c MPa	Internal friction angle φ °
Dry sand	1600	0	30 à 35
Wet sand	1800	0	40
Dry clay	1600	0.02	40 à 45
Wet clay	2000	0.03	15 à 25
Loam	1800	0.025	30 à 45

Table 4.2. Equilibrium load bearing capacity, value of the constants N_p, N_c and N_ρ.

Internal friction angle φ °	N_p · 10^{-4}	N_c · 10^{-4}	N_ρ · 10^{-4}
10	2.47	8.34	0.72
20	6.40	14.83	3.45
30	18.40	30.14	15.19
40	64.19	75.31	81.75

4.2.1.2 Settling load bearing capacity

If the load in the contact between footing and soil increases compared to the value before excavation, settlement occurs. In wet clay and loam, sideway squeezing of ground water makes it a function of time. Initial settlement Δdh in a soil layer of thickness dh yet is given by:

4.2 Performance evaluation

$$\Delta dh = \frac{1}{C_h} \ln\left(\frac{\sigma_1 + \sigma_o}{\sigma_o}\right) dh \tag{4.3}$$

with C_h compression constant at depth h below grade, σ_o the vertical stress component there before excavation started and σ_1 additional vertical stress the building will impose. The compression constant equals:

$$C_h = 0.66\, C_k(h)/\sigma_o \tag{4.4}$$

where $C_k(h)$ is the cone resistance at depth h, measured with a ground-sounding test. A good guess of the additional vertical stress σ_1 is given by Fröhlich's formula:

$$\sigma_1 = \frac{m\,p}{2\pi\,h'^2} \int_A \left[(\cos\alpha)^{m+2}\, dA\right] = \frac{m\,p}{2\pi\,h'^2} \sum_A \left[(\cos\alpha)^{m+2}\, \Delta A\right] \tag{4.5}$$

In it, p is the average load per m² of contact area between footing and soil, h' is depth of a spot below that area and α is the angle between the line through that spot and the vertical in the centre of the contact area. The coefficient m is 3 for clay and 6 for sand. Although the formula shows singularities for h' below 0.4 m, it underscores that surfaces of equal stress under a footing are curved while the stress decreases quickly with depth. Soil settling increases with larger contact area. The building could yet have such stiffness that separate footings act in unity. In that case, the average contact load per m² (p) equals the total building load, divided by the building's footprint.

During excavation, soil decompression obeys an analogous equation as (4.3), however with the decompression constant A_h replacing the compression constant C_h. As the building load increases during construction, soil movement will first follow the decompression curve and then switch to compression once ground pressure exceeds the value before excavating.

That settling increases, the larger the contact area between footing and soil for a given load per m², opposes equilibrium load bearing capacity, where a larger contact area allows more load per m². Figure 4.2 depicts the consequences with equilibrium load bearing capacity determining allowable load at a small footing width, and settling load bearing capacity determining allowable load at a large footing width. At optimum width, both allow identical loads per m².

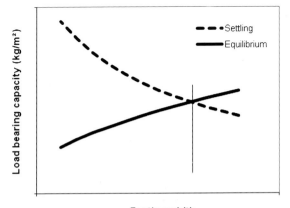

Figure 4.2. Foundation slab, load bearing.

Uniform settling becomes unacceptable when hampering building use. Differential settling demands more attention because it causes toppling, cracking, deformed bays, etc. The rule of thumb is that the tangent of the settling angle between two neighbour footings must not exceed 1/500 (Figure 4.3).

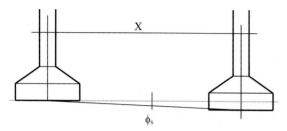

Figure 4.3. Differential settling, angle capacity.

4.2.2 Building physics

Only hygrothermal performances have some importance:

Air-tightness	No requirements
Thermal transmittance	No requirements
Transient response	No requirements
Moisture tolerance	Stable temperature and relative humidity at footing level makes moisture problems unlikely. One exception: historic timber piles. When below the water table for centuries they must remain there. Any table lowering beyond the pile heads causes fast rotting.
Thermal bridging	Must be excluded when constructing refrigerated storage buildings as footings act as thermal bridges, which cause uplifting of the freezing soil below. Also low energy buildings profit from insulated foundations. The lower parts of footings get cellular glass boards, while sides and top may be insulated with cellular glass (if below freezing depth) or XPS. Above the water table, using dense mineral wool is also an option.

4.2.3 Durability

Hygrothermal stress and strain	Stable temperature and relative humidity in the soil at foundation level limits hygrothermal loading. Concrete footings show chemical shrinkage after casting, while organic acids in the soil that diffuse into the concrete may attack it.

4.3 Foundation systems

4.3.1 In general

Selection of a foundation system depends on where below the lowest floor the soil gains enough load-bearing capacity. Choice must be based on a series of cone sounding tests, equally distributed over the building site (Figure 4.4). Foundation design then is an iterative process. In fact, the soil layer thickness needed and its load-bearing capacity not only depends on the building load but also on the foundation's contact area.

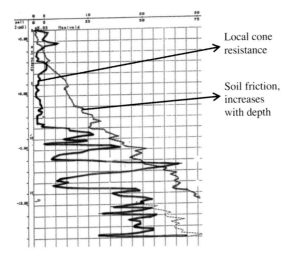

Figure 4.4. Result of a cone-sounding test.

4.3.2 Spread foundations

When thick enough soil layers with the necessary load-bearing capacity sit close below the lowest floor, spread foundations are preferred.

4.3.2.1 Footings

Each load-bearing wall or column gets a footing, wide or large enough to convert the load into the allowable bearing capacity of the soil at the depth considered, see Table 3.3 for estimated bearing capacity values usable as first guess.

Load-bearing walls in single-family houses or apartment buildings with a few floors receive concrete footings. These should be thick enough to spread the wall load under an angle of 30° (Figure 4.5). If calculation demands footings so wide that this 30° rule results in uneconomic footing thicknesses, then reinforced concrete is a cheaper choice. Reinforced concrete footings can be stiff or deformable (Figure 4.5).

Table 4.3. Allowable load bearing capacity.

Soil	(Wet) density kg/m³	Allowable load bearing capacity MPa
Dry sand	1600	0.4 à 0.8
Wet sand	1800	0.2 à 0.4
Dry clay	1600	0.4
Wet clay	2000	0.15
Loam	1800	0.25
Alluvium	1600	0 à 0.015

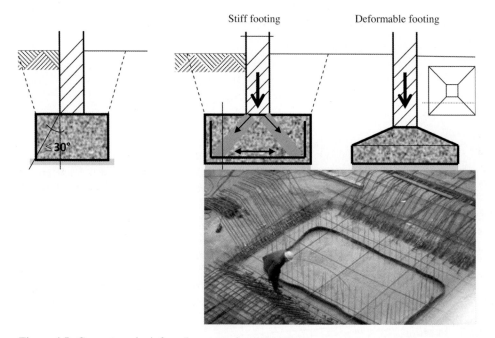

Figure 4.5. Concrete and reinforced concrete footings.

Stiff means a height (h):

$$h \geq (b - b_w)/4 + 0.05 \quad (m) \tag{4.6}$$

where b is footing width and b_w wall width or column mean width. Steel bars at the underside resist the splitting force in the footing.

A deformable footing is lower than the height given by Equation (4.6). It must withstand bending and shear forces. In case differential settling risk is high but load-bearing capacity of the soil is sufficient for footing foundations, they are all coupled with reinforced concrete beams thereby forming a stiff raster.

4.3 Foundation systems

4.3.2.2 Foundation slabs

When footings become so wide they cover a major part of the building footprint or when differential settling must be limited, foundation slabs become the logic choice. Such a slab is thick and monolithic or composed of a beam raster and concrete floor (Figure 4.6). Foundation slabs are typically used for medium and high rises. They are an option for buildings with a basement, even when the load-bearing soil is at two or more meters below grade. Office buildings and high rises sometimes have so many basement floors that a slab remains a valuable choice even with the load bearing soil is 10 to 20 meters below grade.

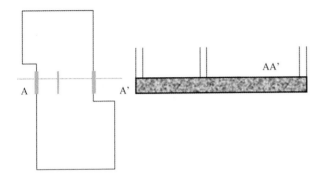

Figure 4.6. Foundation slab.

4.3.2.3 Soil consolidation

In some cases, the cheapest solution consists of consolidating soils with low load-bearing capacity before applying a spread foundation. In alluvial soils, sand piles are an option. In loose sands, vibrating is used to upgrade compactness.

4.3.3 Deep foundations

If soils with good load-bearing capacity are at such a depth that even buildings with basements cannot be founded on footings, if extra basement floors are more expensive or if the soil cannot be ameliorated, then deep founding is the remaining solution.

4.3.3.1 Wells

Well foundations are advisable when the load-bearing soil is 3 to 8 meters below the lowest floor. The wells consist of stacked concrete rings filled with sand or concrete and coupled by a reinforced concrete beam raster (Figure 4.7). After the foundation is marked, one positions the first rings and excavates the soil inside so that the rings sink under their own weight. Once the top of the first ring is at grade level, a second ring is stacked on it and one continues excavating. That way, ring after ring is added until the well reaches the load-bearing soil. Correct load distribution requires that the point of gravity of all wells lie on the same vertical as the point of gravity of the building load.

In massive brick construction, beams beneath all load-bearing walls are calculated on vaulting (Figure 4.8). That makes their section independent of the number of floors. To limit differential settling, the first floors in medium and high-rise buildings are best constructed as stiff reinforced concrete boxes.

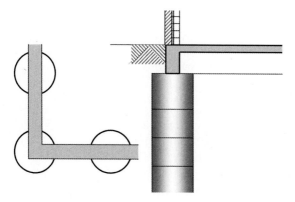

Figure 4.7. Well foundation.

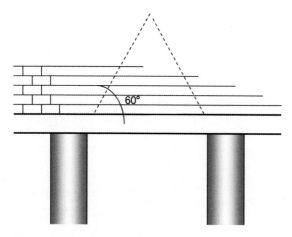

Figure 4.8. Vaulting in masonry.

4.3.3.2 Piles

Piles are preferred when the load bearing soil is 8 meters or more below the lowest floor. Loading is best done axially. For that reason one founds columns on groups of two, three or more piles. That neutralizes errors in driving without bending the piles. After nipping off the pile heads and freeing the anchor bars, each group gets a stiff footing that distributes column load over the piles. Groups of one or two piles are coupled with reinforced concrete beams (Figure 4.9). Pile foundations of high rises are coupled by one large foundation slab. Correct pile distribution dictates that the point of gravity of all piles lies on the same vertical as the point of gravity of the building load. Small horizontal forces compared to the vertical ones are absorbed by loading the piles on shear. In contrast, important horizontal forces demand obliquely driven perimeter piles.

4.4 Specific problems

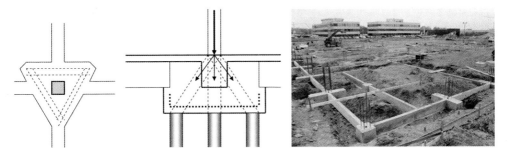

Figure 4.9. Pile foundation.

The number of different pile systems fits quite well in a 2 × 4 array, with the way of driving and pile type as parameters.

Pile type	Way of driving			
	Ramming	Screwing	Vibrating	Press
Formed in the soil	X	X	x	
Prefabricated	X	X	x	x

Driving produces noise and vibrations. For that reason, screwing is preferred on building sites in populated environments. Vibrating gives dynamic nuisance. Pressing requires a counter-weight. Pressed piles apply for reinforcing existing foundations using the building as counter-weight. Prefabricated piles are made of timber, reinforced concrete or steel. With piles formed in the soil, first a steel shaft is driven or screwed into the soil. Once at the correct depth, a steel bar reinforcement is slipped into the shaft, after which stepwise filling with compacted concrete follows while pulling up or unscrewing the shaft out of the ground.

The section fixes the allowable shaft length and the load a pile can bear. For prefabricated reinforced concrete piles with a section of 0.2 × 0.2 m that is ≈ 7 m and ≈ 200 kN, whereas reinforced concrete deep wall piles with a section of 2.3 m^2 can bear 20 000 kN for a shaft length of 50 m. Down drag demands proper consideration, although friction between pile shaft and soil also offers additional load-bearing capacity.

4.4 Specific problems

4.4.1 Eccentrically loaded footings

In skeleton constructions between existing buildings, column footings touching party walls transmit forces quite inefficiently to the ground. The result is a triangular load diagram with the force in the triangle's point of gravity. Perimeter footings are therefore coupled to central footings with reinforced concrete equilibrium beams. That way, the frame, formed by the perimeter column, the equilibrium beam, the central column and the floor beam above delivers the bending force needed to centre the column load on the footing (Figure 4.10).

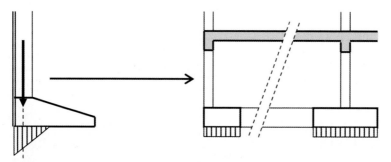

Figure 4.10. Column with eccentrically loaded footing, using equilibrium beams.

4.4.2 Footings under large openings

Massive constructions with footing foundations may have large bays in the facade and inside walls. Many times the height of the below grade wall does not suffice to spread the loads equally over the footing. That way, the non-reinforced concrete has to withstand bending strain, which it cannot, resulting in differential settling of the walls bounding the bay. To avoid this, the footing is stiffened with a reinforced concrete beam so it can withstand bending and shear while the soil below gets equally loaded (Figure 4.11).

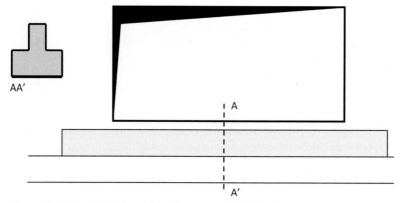

Figure 4.11. Footing below a large bay, use of stiffening beam.

4.4.3 Reinforcing and/or deepening existing foundations

Building construction between existing premises often implies that foundations below party walls have to be reinforced and/or deepened.

4.4.3.1 Footings

The existing foundation is excavated piece-wise in an alternating way so settling risk remains minimal. It is best choice to jump from the middle of the party wall to the corners and then to finish the pieces left between the jumps. Care must be taken that the existing and enlarged footing functions in common, distributing the future larger load equally over the contact surface.

4.4 Specific problems

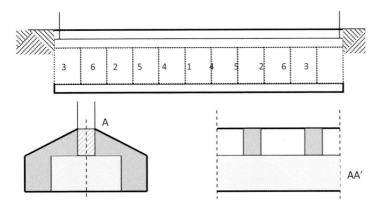

Figure 4.12. Reinforcing existing foundations with enlarged footings.

If deepening is needed, than each piece is excavated to the new depth, the new footing cast and a new wall brick-laid between the new and the existing footing, which one then evens off with the wall's surface. A piece-wise approach is again necessary (Figure 4.12).

4.4.3.2 Wells

Wells only make sense when one has to deepen foundations. Again, the existing footing is excavated piece-wise and cut off. Then the excavation is deepened to allow each concrete ring to be sunk on top of the next. Once at the right depth, one fills the well with concrete and cast part of the supporting beam with lapping bars. Then the next piece is executed, etc. (Figure 4.13).

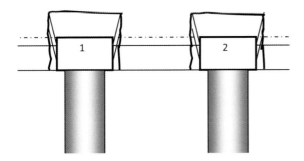

Figure 4.13. Reinforcing existing foundations using wells.

4.4.3.3 Pressed piles

Also here the existing foundation is excavated piecewise and the footing removed. Part of the supporting beam with lapping bars is cast below the founding wall. Once the concrete is strong enough, the beam serves as a support for the hydraulic screw jack that presses the successive pile segments on top of one another into the ground until the right depth is reached. Then one nips the last segment head and uses the space between head and support beam to cast the pile footing.

4.5 References and literature

[4.1] Magnel, G. (1942). *Stabilité des Constructions, Volume 1, 2 et 3*. Rombout-Fecheyr, Editeurs, Gent (in French).

[4.2] Neufert (1964). *Bau-Entwurfslehre*. Ullstein Verlag, 455 pp. (in German).

[4.3] De Beer, E. (1965). *Grondmechanica, delen 1, 2, 3 en 4*. Standaard Wetenschappelijke Uitgeverij (in Dutch).

[4.4] Everard, N., Tanner III, J. (1966). *Reinforced concrete design, Schaum's outline series*. Mc Graw-Hill Book Company.

[4.5] Gerstle, K. (1967). *Basic Structural Design*. Mc Graw-Hill Book Company.

[4.6] Guerrin (1967). *Traité de beton armé 3*. Dunod, Paris (in French).

[4.7] Stichting Bouwresearch (1977). *Funderingen op palen (delen 1, 2 en 3)*. Samson Uitgeverij, Alphen aan de Rijn (in Dutch).

[4.8] WTCB (1983). TV 147: *Funderingen van huizen*. Practische leidraad voor de opvatting en uitvoering van de funderingen van kleine en middelgrote constructies, 68 p. (in Dutch).

[4.9] Frick, Knöll, Neumann, Weinbrenner (1987). *Baukonstruktionslehre, Teil 1 & 2*. B. G. Teubner Verlag, Stuttgart (in German).

[4.10] Oak Ridge National Laboratory (1988). *Building Foundation Design Handbook*. 349 pp.

[4.11] Cziesielski, E. (1990). *Lehrbuch der Hochbaukonstruktionen*. B. G. Teubner Verlag, Stuttgart (in German).

[4.12] Lstiburek, J. (1998). *Builder's Guide*. Building Science Corporation.

[4.13] Hens, H. (2005). *Toegepaste bouwfysica en prestatieanalyse, bouwdelen*. Acco, Leuven (in Dutch).

[4.14] Hens, H. (2010). *Applied Building Physics, Boundary Conditions, Building Performance and Material Properties*. Ernst & Sohn (A Wiley Company), Berlin, 307 p.

[4.15] Hens, H. (2010). *Prestatieanalyse van bouwdelen 1*. Acco, Leuven (in Dutch).

[4.16] ASHRAE (2011). *Handbook of HVAC Applications, Chapter 43*. Atlanta.

5 Building parts on and below grade

5.1 In general

The term on and below grade relates to all building parts other than foundations that demand excavation: walls between grade and footing, crawl spaces, basements and floors on grade. All have their own complexity. Construction of the first three proceeds in excavation with soil stability rendering building more difficult. Heat transfer develops three-dimensionally. Parts below the water table have to withstand water heads.

Chapter five first discusses performance evaluation before analysing construction aspects typical for below and on grade building parts.

5.2 Performance evaluation

5.2.1 Structural integrity

5.2.1.1 Static stability

Foundation walls, crawl spaces and basements transmit and distribute the vertical and horizontal load exerted by the building to and over the foundations. Furthermore, they withstand soil pressure and for parts below the water table, they withstand water heads. On sloped sites, below grade building parts also act as retaining walls, with soil friction and passive soil pressure on the sides away from the slope guaranteeing equilibrium (Figure 5.1). For basements partly below the water table, the weight of the building must compensate upward water pressure. Otherwise, building and basement have to be anchored in the soil with piles (Figure 5.2) or one must ballast the basement, for example by constructing a double floor with gravel in between. In buildings founded on footings, building sections with basement may settle less than those without. Larger soil decompression thanks to deeper excavation and smaller load per m^2 of basement floor are the reasons. A good choice on compressible soil therefore is to found parts outside the basement deeper or to design footings and basements as stiff entities.

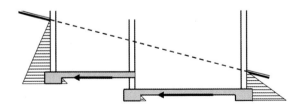

Figure 5.1. Below grade building parts acting as retaining wall.

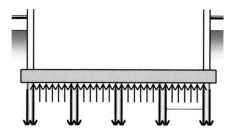

Figure 5.2. Piles used as soil anchor.

5.2.1.2 Strength and stiffness

Basements walls and floors have to withstand axial compression and bending, for the walls by the building above, the soil and, below the water table, by water heads (Figure 5.3), for the floors by the soil, own weight, dead weight, live load and, below the water table, by water heads. In the last case, the lowest basement floor experiences upward water pressure, given by $p = 10\,000\,h$ (N/m^2), where h is the height of the water table above the floor's underside. Greater height quickly increases the pressure. The consequence are field moments in the floor slab with tension in the upper part and support moments below all basement walls with tension in the lower part. Large spans even demand construction as beam raster. Foundation floors finally are subjected to a strong upward bending by soil pressure, induced by the overall building load.

In massive basement walls the axial load usually gains from bending. That keeps the load's eccentricity within the wall's kern, making masonry applicable. Basements including several floors or basements bearing skeleton constructions, however, can experience such large bending moments that reinforced concrete construction is the only way out. In the recent past, precast reinforced concrete basements have taken over an ever-larger market share in residential construction.

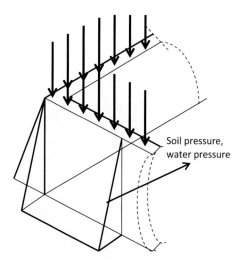

Figure 5.3. Loads on basement walls.

5.2.2 Building physics, heat, air, moisture

5.2.2.1 Air tightness

Problems caused by insufficient air tightness mainly involve ground floors above ventilated crawl spaces. Vents coupling the crawl space to outdoors allow outside air inflow while the draft prone ground floor links the crawlspace to all residential spaces above, which in turn are coupled to the outside through leaks and ventilation grids.

In the case that extract ventilation is applied, air is drawn from the crawl space across the ground floor leaks into the residential space above. That flow is compensated by an identical inflow of outside air into the crawlspace. In other words, floor leaks and vents in the crawl space form a series circuit (Figure 5.4). In case both have the same air permeance exponent b, air leakage can be described as:

$$G_a = \left[\frac{1}{a_{fl}^{1/b}} + 1 \Big/ \left(\sum_{i=1}^{n} a_{i,clsp} \right)^{1/b} \right]^{-b} \Delta P_a^b \tag{5.1}$$

where a_{fl} is the air permeance coefficient of the floor in kg/(s · Pa^b), a_{clsp} the air permeance coefficient of a ventilation opening in the outer wall of the crawl space in kg/(s · Pa^b) and ΔP_a the pressure difference along the path outdoors/crawlspace/living space above in Pa.

Also thermal buoyancy may move air from the crawlspace into the living space above. In windless weather, stack flow equals:

$$G_a = \left[\frac{1}{a_{fl}^{1/b}} + 1 \Big/ \left(\sum_{i=1}^{n} a_{i,clsp} \right)^{1/b} + 1 \Big/ \left(\sum_{j=1}^{m} a_{j,inlet} \right)^{1/b} \right]^{-b} \Delta p_t^b \tag{5.2}$$

where a_{inlet} is the air permeance coefficient of the air inlet grids in the outer wall of the residential space in kg/(s · Pa^b). Thermal stack Δp_t is thereby approximated by:

$$\Delta p_t = 0.043 \left[h_{c,fl} \, \theta_c + h_{fl,rs} \, \theta_i - \left(h_{c,fl} + h_{fl,rs} \right) \theta_e \right] \tag{5.3}$$

with θ_c the temperature in the crawlspace, θ_i the temperature in the living space above, $h_{c,fl}$ the vertical distance between the ventilation openings in the outer crawl space wall and the floor's mid-plane and $h_{fl,rs}$ the vertical distance between the floor's mid-plane and the air inlets in the living space.

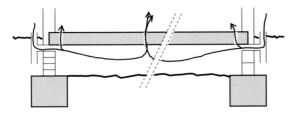

Figure 5.4. Floor leakages between crawlspace and ground floor.

Example

Consider a one storey high detached house with floor area 120 m². On top of the operable windows in the living space, 0.12 m² of air inlet grids are mounted 2 m above the ground floor mid-plane. The crawlspace has 0.4 m² of vents just below the ground floor ($h_{c,fl} = 0.2$ m). Temperatures are 0 °C outside and 20 °C inside while wind velocity is zero. Thermal stack so equals $\approx 0{,}043 \cdot 2.2 \cdot 20 = 1.89$ Pa. Air flow across the air inlet grids and the vents is given by $A\sqrt{2\rho_a/1.5}\,\Delta p_t^{0.5}$ in kg/(s·Pa^b) with ρ_a air density (≈ 1.2 kg/m³), 0.5 the air permeance exponent (b) and $A\sqrt{2\rho_a/1.5}$ the volumetric air permeance coefficients, equal to:

Crawlspace   0.422 m³/(s·Pa^b)
Residential space 0.126 m³/(s·Pa^b)

The volumetric air permeance coefficient of the ground floor is $p\,A/100\sqrt{2/(1.5\rho_a)}$ (m³/(s·Pa^b)), with p the ratio in percentage between floor leakage and total floor area. The air permeance exponent (b) is also 0.5. The resulting airflow in m³/h from crawl space to residential space is shown in Figure 5.5. While negligible below a floor leakage area of 0.002%, airflow quickly increases to stabilize once the leakage area exceeds 0.4%. Of course, that percentage depends on the air permeance of the vents and inlet grids

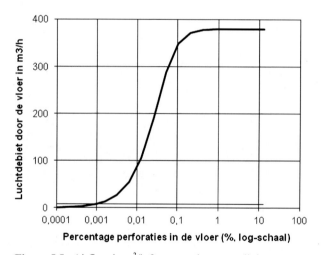

Figure 5.5. Airflow in m³/h from crawl space to living space.
The dashed line shows the maximum leakage value imposed by the Dutch building ordinance.

The consequences of air inflow from the crawl space into the living space can be annoying. In case the crawlspace is moist, considerable water vapour may co-infiltrate. In radon-loaded soils, the same holds for radon. Air from the crawlspace may smell, etc.

Equation (5.2) suggests three measures that limit air inflow: (1) not venting the crawlspace, (2) no air inlets and perfect air-tightness of the residential enclosure, (3) airtight ground floor. The second kills living space ventilation, which is unacceptable. The first demands a warm crawlspace whereas the third seems the most logical choice, though difficult to realize. It is more realistic to limit crawlspace air inflow to a value that avoids mould growth in the residential spaces above. Additional conditions then become a well-insulated envelope ($R_{opaque} \geq 2$ m²·K/W, no problematic thermal bridging), outside air ventilation equal to the value required by law or standard and monthly mean vapour release indoors not exceeding the

5.2 Performance evaluation

indoor climate class 3/4 vapour pressure threshold. For example, the Dutch building ordinance limits air inflow from the crawlspace to 0.072 m³/(m² · h). Implementation in Figure 5.5 shows that value requires a floor leakage ratio below 0.001%.

Where are the leaks in floors above crawlspaces found? Timber decks may have open joints between planks. With prefabricated floors and concrete slabs, heating, water supply pipes, discharge pipes and electrical wiring passages create leaks if not well sealed. Prefabricated floors without concrete topping sometimes suffer from leakage at the supports.

5.2.2.2 Thermal transmittance

For floors on grade, floors above basements, floors above crawlspaces, basement and crawl-space floors and walls in contact with the soil the concept of a 'thermal transmittance' is not applicable anymore. The temperature field in the soil in fact is three-dimensional whereas thermal transmittances presume one-dimensional temperature fields. The concept nevertheless is retained, now called reduced thermal transmittance ($U_{red,fl}$) and given by:

$$U_{red,fl} = a\, U_{o,fl} \tag{5.4}$$

with 'a' a reduction factor and $U_{o,fl}$ the thermal transmittance as if the underside of a floor on grade, a heated basement floor and its outer walls were facing the outside environment and, as if the underside of a floor above unheated basements and crawlspaces were facing an inside environment. The problem then becomes to calculate that factor.

Software

The best method is use software. The tools for three-dimensional heat transport in soils with constant thermal conductivity are quite simple. Figure 5.6 shows the results for a heated basement with a 25 cm thick concrete floor and 20 cm thick concrete walls. Thermal conductivity of the concrete is 2.2 W/(m · K), of the soil 2 W/(m · K). Reduction factors are 0.077 for the floor and 0.23 for the walls. When the floor and walls consisted of light-weight concrete, $\lambda = 1$ W/(m · K), the values should have been 0.088 for the floor and 0.29 for the walls. Or, reduction decreases with better insulation of the basement floor and walls.

A better choice is to use software that considers combined heat and moisture transfer in the soil. That way, enthalpy displacement is included. Figure 5.7 gives an example. With a thermal transmittance of 0.7 W/(m² · K) for the basement floor and the walls, reduction factors reach 0.26 and 0.62 respectively.

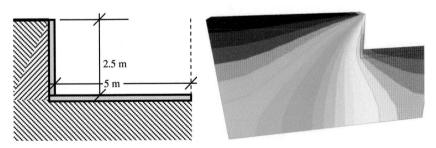

Figure 5.6. Heated basement with concrete walls and floor (20 °C inside, 0 °C outside), isotherms in the soil (calculated with software for steady state three-dimensional heat transport, the soil with thermal conductivity 2 W/(m · K)).

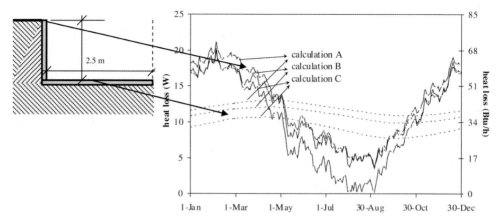

Figure 5.7. Heated basement (20 °C inside, 0 °C outside), heat losses to the soil. A is the reference. In B moisture content in the soil equals the annual mean. C considers the equivalent temperature outdoors. The curves with highest amplitude represent heat flow across the walls, the other heat flow across the floor.

Standard EN ISO 13370

Floor on grade

Calculation starts with the characteristic floor dimension:

$$B' = 2\, A_\text{fl}/P \quad \text{(m)} \tag{5.5}$$

where A_fl is floor area and P the part of the floor perimeter facing the outdoors, called the free perimeter (Figure 5.8). The floor is then replaced by an equivalent soil thickness (d_t):

$$d_\text{t} = d_\text{fw} + \lambda_\text{gr}\left(\frac{1}{h_\text{e}} + R_{\text{T,fl}} + \frac{1}{h_\text{i}}\right) \quad \text{(m)} \tag{5.6}$$

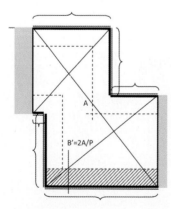

Figure 5.8. Floor on grade, characteristic dimension. Braces show the part of the perimeter facing outdoors, while the hatched surface gives the horizontal width of the horizontal perimeter insulation.

5.2 Performance evaluation

where d_{fw} is the average width in m of the foundation walls under the free perimeter, λ_{gr} the thermal conductivity of the soil in W/(m · K), $R_{T,fl}$ the thermal resistance of the floor in m² · K/W, h_i the surface film coefficient indoors, 6 W/(m² · K)), and h_e the surface film coefficient outdoors, 25 W/(m² · K). If thermal conductivity of the soil is unknown, the following values are proposed:

Soil	Thermal conductivity W/(m · K)	Volumetric heat capacity J/(m³ · K)
Clay or silt	1.5	$3 \cdot 10^6$
Sand or gravel	2.0	$2 \cdot 10^6$
Homogeneous rock	3.5	$2 \cdot 10^6$

Finally, the reduction factor is calculated, its value depending on the ratio between the equivalent soil thickness and the characteristic floor dimension (see Figure 5.9):

$$d_t < B': \ a = \frac{1}{U_{o,fl}} \left(\frac{2 \lambda_{gr}}{\pi B' + d_t} \right) \ln \left(\frac{\pi B'}{d_t} + 1 \right)$$

$$d_t \geq B': \ a = \frac{1}{U_{o,fl}} \left(\frac{\lambda_{gr}}{0.457 B' + d_t} \right) \tag{5.7}$$

Without perimeter insulation, calculation comes to a halt. If perimeter insulation is applied, reduced thermal transmittance turns into:

$$U_{red,fl} = a \, U_{o,fl} + 2 \frac{\psi}{B'} \quad (W/(m^2 \cdot K)) \tag{5.8}$$

with ψ a negative linear thermal transmittance along the free perimeter, a value depending on the equivalent soil thickness of the perimeter isolation:

$$d' = \lambda_{gr} \left(R_{ins} - \frac{d_{ins}}{\lambda_{gr}} \right) \quad (m) \tag{5.9}$$

and given by:

$$\psi = -\frac{\lambda_{gr}}{\pi} \left[\ln \left(\frac{D}{d_t} + 1 \right) - \ln \left(\frac{D}{d_t + d'} + 1 \right) \right] \quad (W/(m \cdot K)) \tag{5.10}$$

where D is the horizontal width of the insulation strip and d_t the equivalent soil thickness of the floor (Figure 5.8). In case the perimeter insulation is lined vertically against the foundation walls or if these walls are constructed of insulating blocks (R_{ins} and d_{ins} in [5.9]), linear thermal transmittance in terms of mean thermal resistance and thickness of the foundation walls then becomes:

$$\psi = -\frac{\lambda_{gr}}{\pi} \left[\ln \left(\frac{2H}{d_t} + 1 \right) - \ln \left(\frac{2H}{d_t + d'} + 1 \right) \right] \tag{5.11}$$

with H the height of the perimeter insulation or height of the foundation walls.

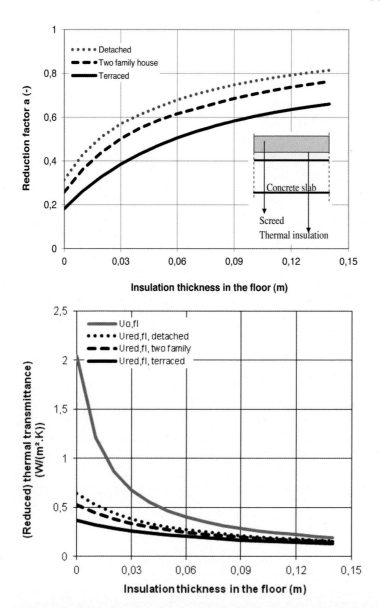

Figure 5.9. Floor on grade, reduction factor and reduced thermal transmittance (EN ISO 13370). Area 77.8 m², free perimeter 36 m if detached, 25.2 m if two family and 14.4 m if terraced. Thermal resistance without insulation 0.32 m² · K/W, thermal conductivity insulation 0.03 W/(m · K), soil 2.0 W/(m · K).

5.2 Performance evaluation

Heated basement or heated space partly below grade

First, the characteristic dimension of the basement or space below grade is calculated with the facade length as free perimeter. The reduction factor for the floor then is:

$$d_t + H/2 < B': \quad a_{fl} = \frac{1}{U_{o,fl}} \left[\frac{2\lambda_{gr}}{\pi B' + d_t + H/2} \ln\left(\frac{\pi B'}{d_t + H/2} + 1\right) \right] \quad (5.12)$$

$$d_t + H/2 \geq B': \quad a_{fl} = \frac{1}{U_{o,fl}} \left(\frac{\lambda_{gr}}{0.457 B' + d_t + H/2} \right) \quad (5.13)$$

with B' the (mean) equivalent soil thickness of the outer walls (see [5.5]), H the mean height between the floor's underside and grade and d_t the equivalent soil thickness of the floor.
For the walls along the free perimeter, the equivalent soil thickness intervenes:

$$d_{t,w} = \lambda_{gr} \left(\frac{1}{h_e} + R_{T,w} + \frac{1}{h_i} \right) \quad (5.14)$$

giving as a reduction factor:

$$d_{t,w} \geq d_t: \quad d_t + H/2 \geq B' \quad a_w = \frac{1}{U_{o,w}} \left[\frac{2\lambda_{gr}}{\pi H} \left(1 + \frac{0.5 d_t}{d_t + H}\right) \ln\left(\frac{H}{d_{t,w}} + 1\right) \right] \quad (5.15)$$

$$d_{t,w} < d_t: \quad a_w = \frac{1}{U_{o,w}} \left[\frac{2\lambda_{gr}}{\pi H} \left(1 + \frac{0.5 d_{t,w}}{d_{t,w} + H}\right) \ln\left(\frac{H}{d_{t,w}} + 1\right) \right]$$

The reduced heat transmission coefficient of a heated basement then looks like:

$$H_{red} = a_{fl} U_{o,fl} A_{fl} + a_w U_{o,w} P H \quad (5.16)$$

The free perimeter (P) and floor area (A_{fl}) are measured out to out.

Floor above crawlspace

The reduction factor is given by:

$$a_{red} = \frac{1}{U_{o,fl}} \left(\frac{1}{U_{o,fl}} + \frac{1}{U_{red,gr} + U_x} \right)^{-1} \quad (5.17)$$

with $U_{o,fl}$ the thermal transmittance of the floor (calculated with $h_i = h_e = 6$ W/(m² · K)), $U_{red,gr}$ the reduced thermal transmittance of the bottom and U_x a fictitious thermal transmittance combining crawlspace ventilation with the heat flow across the above grade crawlspace outer walls.

Crawlspace bottom

Characteristic dimension B' of the crawlspace bottom is:

$$B' = 2\, A_{\text{clsp}}/P$$

with A_{clsp} bottom area out to out and P free perimeter. The equivalent soil thickness of the bottom is given by ($h_i = 6$ W/(m$^2 \cdot$ K) and $h_e = 25$ W/(m$^2 \cdot$ K)):

$$d_{\text{gr}} = d_{\text{fw}} + \lambda_{\text{gr}}\left(\frac{1}{h_i} + R_{\text{T,clsp}} + \frac{1}{h_e}\right)$$

where d_{fw} is the thickness of the foundation walls in m and $R_{\text{T,clsp}}$ the thermal resistance of the bottom slab. The bottom's reduced thermal transmittance then becomes:

$$U_{\text{red,gr}} = \frac{2\lambda_{\text{gr}}}{\pi B' + d_{\text{gr}}} \ln\left(\frac{\pi B'}{d_{\text{gr}}} + 1\right)$$

Crawlspace ventilation and the heat flow across the above grade crawlspace walls

The fictitious thermal transmittance is calculated as:

$$U_x = \frac{2\,H\,U_w}{B'} + 1450\, A_{\text{vent}}\, \frac{v_w\, f_w}{B'}$$

In that formula H is the mean wall height between grade and the underside of the ground floor, U_w is that wall's thermal transmittance, A_{vent} is the area of vents per meter run along the crawlspace's free perimeter in m^2/m, v_w the annual mean wind speed measured at the nearest weather station in m/s, and f_w the wind shielding factor, equal to:

Situation	Example	wind shielding factor f_w
Sheltered	City centre	0.02
Average	Suburban, village	0.05
Exposed	Rural, open	0.10

Floor above unheated basement

The reduction factor equals:

$$a_{\text{redl}} = \frac{1}{U_{\text{o,fl}}}\left\{\frac{1}{U_{\text{o,fl}}} + \frac{A_{\text{bas}}}{A_{\text{bas}}\,U_{\text{red,fl,bas}} + P\left[H\,U_{\text{red,w,bas}} + (H_{\text{bas}} - H)\,U_{\text{w,bas}}\right] + 0.33\,n\,V}\right\}^{-1} \quad (5.18)$$

where A_{bas} is the basement area, H_{bas} the height between the underside of the basement floor and the underside of the floor above, H the height between the underside of the basement floor and grade, P the free perimeter, $U_{\text{o,fl}}$ the thermal transmittance of the floor above (calculated with $h_i = h_e = 6$ W/(m$^2 \cdot$ K)), $U_{\text{red,fl,bas}}$ reduced thermal transmittance of the basement floor, $U_{\text{red,w,bas}}$ the thermal transmittance of the basement outer walls, $U_{\text{w,bas}}$ the thermal transmittance of the

5.2 Performance evaluation

basement outer walls above grade (calculated with $h_i = 8$ W/(m² · K) and $h_e = 25$ W/(m² · K)), n the ventilation rate in the basement and V the basement volume out to out.

Basement floor

The characteristic dimension B' is $2\,A_{bas}/P$. Equivalent soil thickness becomes ($h_i = 6$ W/(m² · K) and $h_e = 25$ W/(m² · K)):

$$d_t = d_{w,bas} + \lambda_{gr}\left(\frac{1}{h_i} + R_{T,fl,bas} + \frac{1}{h_e}\right)$$

with $d_{w,bas}$ the thickness of the basement outer walls in m, and $R_{T,fl,bas}$ the thermal resistance of the basement floor. Reduced thermal transmittance calculates as:

$$d_t + H/2 < B': \quad U_{red,fl,bas} = \frac{2\,\lambda_{gr}}{\pi\,B + d_t + H/2}\ln\left(\frac{\pi\,B'}{d_t + H/2} + 1\right)$$

$$d_t + H/2 \geq B': \quad U_{red,fl,bas} = \frac{\lambda_{gr}}{0.457\,B' + d_t + H/2}$$

Below grade basement outer walls

Equivalent soil thickness is ($h_i = 6$ W/(m² · K) and $h_e = 25$ W/(m² · K)):

$$d_{t,w} = \lambda_{gr}\left(\frac{1}{h_i} + R_{T,w,bas} + \frac{1}{h_e}\right)$$

with $R_{T,w,bas}$ the thermal resistance of the basement floor. Reduced thermal transmittance becomes:

$$d_{t,w} \geq d_t: \quad U_{red,w,bas} = \left[\frac{2\,\lambda_{gr}}{\pi\,H}\left(1 + \frac{0.5\,d_t}{d_t + H}\right)\ln\left(\frac{H}{d_{t,w}} + 1\right)\right]$$

$$d_{t,w} < d_t: \quad U_{red,w,bas} = \left[\frac{2\,\lambda_{gr}}{\pi\,H}\left(1 + \frac{0.5\,d_{t,w}}{d_{t,w} + H}\right)\ln\left(\frac{H}{d_{t,w}} + 1\right)\right]$$

Method of the perimeter circles

Although a few meters below any excavation soil temperature equals the annual mean outdoors, EN ISO 13370 does not consider that when calculating reduction factors and reduced thermal transmittances. It only does so when stepping to heat fluxes as reduced thermal transmittances are multiplied with the difference between the temperature indoors and the annual mean temperature outdoors. However, when looking at the isoflux lines in the soil, a clear difference is noticed between a perimeter zone where the lines form circles from indoors to grade and a central zone, where the lines plunge vertically into the soil (Figure 5.10).

That allows splitting the heat flow in two parts: a flow straight to the isotherm in the soil some 5 meters below the excavated volume and a circular perimeter flow:

$$\Phi_{CT} = \psi_{per}\,P\,(\theta_i - \theta_e) + U_c\,r_{fl}\,A_{fl}\,(\theta_i - \theta_{em}) \tag{5.19}$$

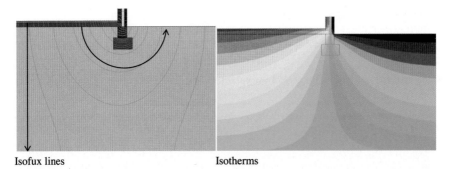

Isofux lines   Isotherms

Figure 5.10. Slab on grade: the isoflux lines along the perimeter circle from indoors to outdoors.

In that formula, θ_{em} is the annual mean temperature outdoors, A_{fl} the floor area out to out, r_{fl} a reduction factor accounting for the active floor area being smaller than total floor area, and P the free perimeter. U_c is thermal transmittance of floor plus soil down to the isotherm and ψ_{per} is a linear thermal transmittance along the free perimeter. The reduction factor thus depends on outside temperature:

$$a = \frac{1}{U_{o,fl}} \left(\psi_{per} \frac{P}{A_{fl}} + r_{fl} U_c \frac{\theta_i - \theta_{em}}{\theta_i - \theta_e} \right) \tag{5.20}$$

The method allows solving any below grade construction. For a crawlspace or basement, the split between perimeter heat flow and soil flow coincides with the line on the excavated outer walls where circles up and circles plus vertical down into the soil gave identical flows. A situation below the water table is easily solved by adapting thermal conductivity of the soil.

Floor on grade

Heat flow at the free perimeter develops along two-quarter circles, one between floor and foundation wall and a second between foundation wall and grade (Figure 5.11). The largest radius (r_{max}) stands for flow rate along the perimeter equal to the vertical flow rate into the soil:

$$r_{max} = \frac{\lambda_{gr}}{\pi} \left[\frac{(\theta_i - \theta_e)}{U_c (\theta_i - \theta_{e,m})} - (R_{T,fl} + R_{T,fw} + 0.21) \right]$$

with $R_{T,fl}$ the thermal resistance surface to surface of the floor and $R_{T,fw}$ the thermal resistance surface to the surface of the foundation wall.

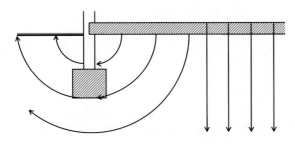

Figure 5.11. Slab on grade: heat loss, perimeter circles method.

5.2 Performance evaluation

Central thermal transmittance U_c becomes:

$$U_c = \frac{1}{1/h_i + R_{T,fl} + 5/\lambda_{gr}} \tag{5.21}$$

The elementary heat flow per meter run of free perimeter is:

$$d\Phi = \frac{(\theta_i - \theta_e) \, dr}{0.21 + R_{T,fl} + R_{T,fw} + \pi r/\lambda_{gr}}$$

With $0.21 + R_{T,fl} + R_{T,fw}$ as α_1 and π/λ_{gr} as α_2, the solution of that differential equation is (integration between $0 \le r \le r_{max}$) $\Phi = (\theta_i - \theta_e) [\ln(1 + \alpha_2 r_{max}/\alpha_1)] / \alpha_2$, giving as linear thermal transmittance:

$$\psi_{per} = \frac{1}{\alpha_2} \ln\left(1 + \frac{\alpha_2 r_{max}}{\alpha_1}\right) \tag{5.22}$$

Floor insulation is embedded in the total thermal resistance $R_{T,fl}$, and perimeter insulation in the total thermal resistance $R_{T,fw}$. Floor area reduction factor r_{fl} equals: $1 - P(r_{max} + d_w) / A_{fl}$ with P the free perimeter and d the thickness of the foundation walls. Entering Equations (5.21), (5.22) and r_{fl} into (5.20) gives the reduction factor for a floor on grade.

Heated basement or heated space partly below grade

Three heat flows now intervene: one between the outer walls below grade and grade, one between these walls and the isotherm 5 meters below the basement floor and one between that floor and that isotherm (Figure 5.12). The first heat flow develops along quarter circles with the largest radius such that the flow rate to grade and the one to the isotherm in the soil are equal. The elementary heat flow per meter free perimeter becomes:

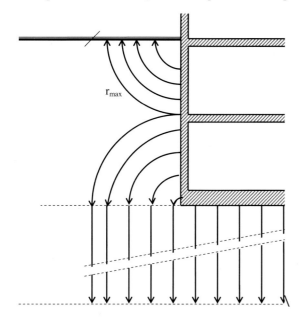

Figure 5.12. Heated basement, heat loss, perimeter circles method.

$$d\Phi_1 = \frac{(\theta_i - \theta_e)\, dr}{\underbrace{0.17 + R_{T,w}}_{\alpha_3} + \underbrace{\pi r/(2\lambda_{gr})}_{\alpha_4}} = \frac{(\theta_i - \theta_e)\, dr}{\alpha_3 + \alpha_4\, r}$$

Solving gives: $\Phi_1 = (\theta_i - \theta_e)\left[\ln(1 + \alpha_4\, r_{max}/\alpha_3)\right]/\alpha_4$ or:

$$\psi_{per,1} = \frac{1}{\alpha_4}\ln\left(1 + \frac{\alpha_4\, r_{max}}{\alpha_3}\right) \tag{5.23}$$

The thermal resistance $R_{T,w}$ includes perimeter insulation.

The second heat flow follows in quarter circles to a level coinciding with the underside of the lowest basement floor, the largest having as a radius the height of the basement outer wall below grade minus r_{max}, to plunge from there vertically to the isotherm 5 meters below. The elementary heat flow per meter free perimeter becomes:

$$d\Phi = \frac{(\theta_i - \theta_{em})\, dr}{\underbrace{0.13 + R_{T,w} + 5/\lambda_{gr}}_{\alpha_5} + \underbrace{\pi r/(2\lambda_{gr})}_{\alpha_4}} = \frac{(\theta_i - \theta_{em})\, dr}{\alpha_5 + \alpha_4\, r}$$

Solution: $\Phi_2 = (\theta_i - \theta_{em})\left[\ln(1 + \alpha_4\, r_2/\alpha_5)\right]/\alpha_4$, or:

$$\psi_{per,2} = \frac{1}{\alpha_4}\ln\left(1 + \frac{\alpha_4\, r_2}{\alpha_3}\right) \tag{5.24}$$

Total linear thermal transmittance per meter free perimeter then becomes:

$$\psi_{per} = \psi_{per,1} + \psi_{per,2}\left(\frac{\theta_i - \theta_e}{\theta_i - \theta_{em}}\right) \tag{5.25}$$

The third heat flow develops between the basement floor and the isotherm 5 meters below. Thermal transmittance is:

$$U_c = \frac{1}{0.17 + R_{T,fl} + 5/\lambda_{gr}} \tag{5.26}$$

The reduced transmission coefficient for a heated basement or for the parts below grade of a heated space then equals:

$$H = \psi_{per}\, P + r_{fl}\, U_c\, A_{fl}\, \frac{\theta_i - \theta_{em}}{\theta_i - \theta_e}$$

Floor above crawlspace: heat balance method

A simplified calculation of the reduction factor is based on solving a system of three heat balances, one for the floor's underside, one for the crawlspace bottom and one for the air in the crawlspace. Not accounted for is radiation with the crawlspace walls. Due to varying temperature along their height, the mathematics are too complex to describe that exchange and the view factor with the floor above is mostly too small to have significant impact.

5.2 Performance evaluation

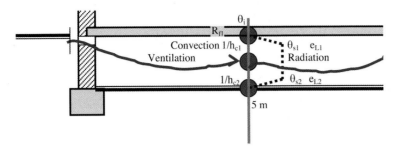

Figure 5.13. Floor above crawlspace: method of the heat balances.

The annual isotherm is located 5 m below the bottom floor. The mean operative temperature in the inhabited space above and the floor's underside temperature (θ_{s1}) form the driving force for a steady state heat flow across the floor above. That surface temperature in turn follows from the balance between the heat flow across, the underside convection with the crawlspace air (θ_c) and radiant exchange with the bottom (θ_{s2}). The heat balances to be solved are then:

Floor
$$\frac{\theta_i - \theta_{s1}}{R_\mathrm{fl} + 1/6} + h_{c1}(\theta_c - \theta_{s1}) + \frac{C_b F_T (\theta_{s2} - \theta_{s1})}{1/e_{L1} + 1/e_{L2} - 1} = 0$$

Bottom
$$\frac{9.8 - \theta_{s2}}{3/\lambda_b} + h_{c1}(\theta_c - \theta_{s2}) + \frac{C_b F_T (\theta_{s1} - \theta_{s2})}{1/e_{L1} + 1/e_{L2} - 1} = 0$$

Air
$$\left[h_{c1}(\theta_{s1} - \theta_c) + h_{c1}(\theta_{s2} - \theta_c) \right] A_\mathrm{fl}$$
$$+ P(\psi_\mathrm{per} + U_{c,w}\, z)(\theta_c - \theta_c) + \rho_a c_a n V_c (\theta_e - \theta_c) = 0$$

Figure 5.13 clarifies the symbols. ψ_per is the linear thermal transmittance along the free perimeter representing that part of the crawlspace walls contacting the soil, z is the height between grade and the slab above and $U_{c,w}$ the thermal transmittance of those parts of the crawlspace outer walls facing the outside air.

Calculating the linear thermal transmittance ψ_per resembles that of walls contacting the soil, however with the maximum radius now equal to the height from the underside of the bottom floor to grade.

The solution for the reduction factor and reduced thermal transmittance for a floor above crawlspace so becomes:

$$a = \frac{U_\mathrm{red,fl}}{U_\mathrm{o,fl}} = \frac{R_{T,\mathrm{fl}} + 2/6}{R_{T,\mathrm{fl}} + 2/6} \left(\frac{\theta_i - \theta_{s1}}{\theta_i - \theta_e} \right) \qquad U_\mathrm{red,fl} = U_\mathrm{o,fl} \left(\frac{\theta_i - \theta_{s1}}{\theta_i - \theta_e} \right) \qquad (5.27)$$

Due to the large thermal inertia of crawlspaces, only annual and monthly means should be considered. Figure 5.14 gives annual results for a floor above a 60 cm high crawlspace under a terraced house with a ground floor of 7.2 × 10.8 m², a free perimeter of 14.4 m and floor insulation between concrete deck and screed. Indoors, 21 °C is assumed. The figure also contains EN ISO 13370 values for 0.003 m² ventilation openings per meter run of free perimeter.

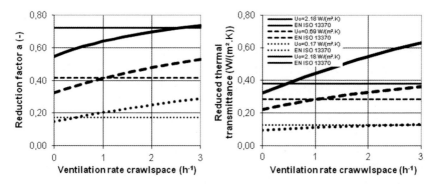

Figure 5.14. Annual mean reduction factor and reduced thermal transmittance for a floor above a 60 cm high crawlspace under a terraced house (ground floor 7.2 × 10.8 m², free perimeter 14.4 m).

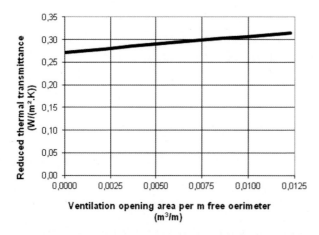

Figure 5.15. Annual mean reduced thermal transmittance according to EN ISO 13370.

Both methods show reduction decreases and reduction factor increases with larger insulation thickness in the floor above. Floor insulation is therefore less efficient than facade or roof insulation. At first sight, according to EN ISO 13370, crawlspace venting has no impact, whereas the heat balance method predicts less reduction with increasing ventilation. This of course is a distorted conclusion as more venting in EN ISO 13370 means more ventilation openings as Figure 5.15 proves. Even more, higher ventilation rates presume larger crawlspace vents at given wind speed. Or, both methods produce analogous conclusions about ventilation: decrease in reduction and increase in reduction factor.

Table 5.1 gives the monthly mean reduction factor and reduced thermal transmittance of the given floor for the reference year at Uccle, Belgium. Reduction factors in summer look much higher than in winter. This is a direct consequence of the constant soil temperature 5 meters below the crawlspace, in Uccle 10.3 °C.

An alternative for a ventilated crawlspace is a warm crawlspace. In this case, instead of the floor above, the crawlspace bottom and outer walls are insulated, while venting is omitted (Figure 5.16). Column 7 and 8 in Table 5.1 give the monthly mean reduction factor and reduced thermal transmittance in the case that both get 6 cm XPS. The table shows that warm crawlspaces give very low reduction factors during the heating season.

5.2 Performance evaluation

Table 5.1. Floor above crawlspace, monthly mean reduction factor at Uccle, Belgium ($n = 1\ h^{-1}$, warm $n = 0\ h^{-1}$, temperatures in bold: heating season).

Month	Temp °C	Insulated floor above $U_{o,fl} = 0.67$ W/(m² · K)		Insulated floor above $U_{o,fl} = 0.17$ W/(m² · K)		Warm crawlspace 6 cm XPS	
		a	$U_{red,fl}$	a	$U_{red,fl}$	a	U_{red}
EN ISO 13370		0.42	0.29	0.72	0.13	0.09	0.20
J	3.2	0.41	0.27	0.64	0.11	0.07	0.16
F	3.9	0.42	0.28	0.65	0.11	0.07	0.16
M	5.9	0.44	0.29	0.69	0.12	0.08	0.18
A	9.2	0.50	0.34	0.77	0.14	0.10	0.22
M	**13.3**	**0.64**	**0.44**	**0.98**	**0.17**	**0.14**	**0.31**
J	**16.2**	**0.89**	**0.60**	**1.37**	**0.24**	**0.22**	**0.49**
J	**17.6**	**1.13**	**0.76**	**1.75**	**0.31**	**0.30**	**0.66**
A	**17.6**	**1.13**	**0.76**	**1.75**	**0.31**	**0.30**	**0.66**
S	**15.2**	**0.76**	**0.51**	**1.18**	**0.21**	**0.18**	**0.40**
O	11.2	0.55	0.37	0.85	0.15	0.12	0.25
N	6.3	0.45	0.30	0.70	0.12	0.08	0.18
D	3.5	0.42	0.28	0.64	0.11	0.07	0.16

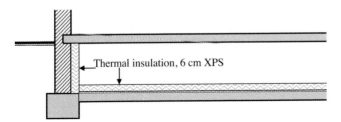

Figure 5.16. Warm crawlspace.

Floors above unheated basements: heat balance method

The geometry of basements makes splitting between convection and radiation too complex, which is why a simple heat balance based on operative temperatures is used (Figure 5.17):

$$A_{bas}\left[U_{o,fl}\left(\theta_i - \theta_{bas}\right) + U_{fl,bas}\left(10.3 - \theta_{bas}\right)\right]$$
$$+ P\left(\psi_{per} + z\, U_{bas,w}\right)\left(\theta_e - \theta_{bas}\right) + 0.8\, \rho_a\, c_a\, n\, V_{bas}\left(\theta_e - \theta_{bas}\right) + \Phi_{F,bas} = 0 \quad (5.28)$$

In that equation $U_{o,fl}$ is the thermal transmittance of the floor above, θ_{bas} the operative temperature in the basement, $U_{fl,bas}$ the thermal transmittance of the basement floor including the 5 meter thick soil layer below, ψ_{per} the linear thermal transmittance, representing the basement outer walls below grade, now with the maximum isoflux line radius equal to the wall's height

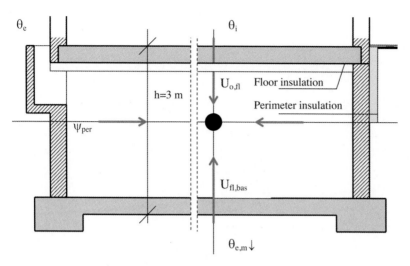

Figure 5.17. Basement, heat balance method.

till grade, z the height between grade and the slab above, $U_{bas,w}$ the thermal transmittance of that part of the basement walls, V_{bas} the air volume in the basement, A_{bas} the basement area measured out to out and Φ_F internal heat gains in the basement. The reduction factor for the floor above and the reduced thermal transmittance follow from:

$$a = \frac{U_{red,fl}}{U_{o,fl}} = \frac{R_{T,fl} + 2/6}{R_{T,fl} + 2/6} \left(\frac{\theta_i - \theta_{sl}}{\theta_i - \theta_e} \right) \qquad U_{red,fl} = U_{o,fl} \left(\frac{\theta_i - \theta_{sl}}{\theta_i - \theta_e} \right)$$

Table 5.2 resumes the crawlspace example, now for a basement. The table gives the heating season mean reduction factor for a well and non-insulated floor above. The basement has perimeter insulation up to 1 meter below grade. The reduction factor increases with better floor insulation and more ventilation. Again, floor insulation sees its efficiency dropping compared to roof and facade insulation. Limited ventilation gives, as Figure 5.18 shows, rather high winter and quite low summer temperatures in the basement.

Table 5.2. Terraced dwelling, ground floor area out to out 7.2 × 10.8 m², free perimeter 14.4 m, $\theta_i = 21$ °C, floor above basement. Heating season mean reduction factor for Uccle, Belgium (October till April, Table 5.1).

Ventilation (h⁻¹)	Heating season mean reduction factor		
Floor $U_{o,fl}=$	Not insulated 2.22 W/(m²·K)	Insulated 0.69 W/(m²·K)	Insulated 0.17 W/(m²·K)
0	0.25	0.49	0.74
1	0.42	0.68	0.86
2	0.54	0.77	0.91
4	0.67	0.86	0.94

5.2 Performance evaluation

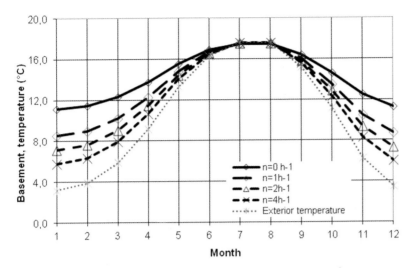

Figure 5.18. Terraced dwelling, ground floor out to out 7.2 × 10.4 m², monthly mean basement temperature, impact of basement ventilation.

Performance requirements

As listed in Table 5.3, most European countries have legal thermal insulation requirements for on and below grade building parts. That is done by fixing a minimum for the thermal resistance of the floor assembly and/or a maximum for the annual mean thermal transmittance (U_{red}), calculated according to EN ISO 13370.

Table 5.3. Floors on grade, reduced thermal transmittance or thermal transmittance and reduction factor, requirements in different European countries.

Country	R_T m²·K/W	U_{red} W/(m²·K)	U_o W/(m²·K)	a
Belgium	1.00 (2006) 1.30 (2012) 1.75 (2014)	0.40 (2006) 0.35 (2012) 0.30 (2014)		
Denmark		0.20–0.30		
Germany			0.30	0.5
France		0.47		
The Netherlands	2.5			
Norway		0.15		
UK		0.25		
Finland		0.25		
Sweden[1]			0.15	0.75

[1] In Sweden, severe requirements are imposed for the mean thermal transmittance of the envelope, not for the thermal transmittances of separate building parts

Table 5.4. Detached dwelling, floor on grade with thermal resistance 1.75 m² · K/W, reduced thermal transmittance as function of its characteristic dimension.

Characteristic dimension m	$U_{red,fl}$ W/(m² · K)
4	0,36
6	0,32
8	0,28
10	0,25

In case a minimum thermal resistance and a maximum thermal transmittance are forwarded, the most severe of both requirements should be met. Take for example as mandatory a thermal resistance 1.75 m² · K/W. For a building on sandy soil, reduced thermal transmittance equals the values of Table 5.4, showing the double requirement ($R_T \geq 1.75$ m² · K/W, $U_{red} \leq 0.3$ W/(m² · K)) is only met for floors with a characteristic dimension above 7 m.

Floor assemblies

Although the insulation of floors on grade, floors above crawlspaces and floors above unheated basements is part of 'design and execution', we introduce it here. There are three ways of doing it (Figure 5.19): (1) casting the screed in well-insulating low-density concrete, (2) laying the insulation below the slab or fixing it against the underside of the floor slab or (3) putting it between a levelled slab and the screed. With crawlspaces, alternative four is the warm crawlspace, not ventilated, with insulation at the bottom floor and against the outer walls (Figure 5.16). Whether these four possibilities require additional layers to guarantee correct hygrothermal response will be discussed in the performance checks that follow.

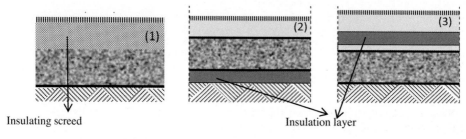

Insulating screed   Insulation layer

Figure 5.19. Floor assemblies.

5.2.2.3 Transient response

Thanks to the ground around, below grade building parts show remarkably high thermal inertia. Table 5.5 list the admittances and time shifts on a daily, weekly and annual basis for a 29 cm thick masonry basement wall and a concrete basement floor. Even for a one-year temperature swing both admittances still pass thermal transmittance with time shifts between 16 and 20 days. Alternately, a correct picture of the thermal response of a basement presumes treating even annual temperature swings as transient.

5.2 Performance evaluation

Table 5.5. Thermal admittance and time shift of basement walls and floor.

Assembly, U_o in W/(m²·K)		Admittance (Ad) in W/(m²·K)						Steady state
				Time shift in hours				
Period		1 day		1 week		1 year		
29 cm thick masonry wall				U_{red} (W/(m²·K))				
No insulation	$U_{o,w} = 2.2$ W/(m²·K)	4.8	1.3	2.7	13.5	0.60	432	0.56
Perimeter insulation	$U_{o,w} = 0.6$ W/(m²·K)	4.8	1.3	2.9	13.5	0.33	469	0.31
Concrete floor								
No insulation	$U_{o,fl} = 4.5$ W/(m²·K)	5.0	0.7	3.4	11.7	0.71	428	0.64
Insulation	$U_{o,fl} = 0.6$ W/(m²·K)	5.2	0.9	2.6	23.8	0.34	377	0.34

Transient response is taken into account by splitting heat flows in an annual mean, calculated with CVM, FEM, EN ISO 13370 or the perimeter circles and an annual amplitude with time shift, calculated using FEM, CVM or the formulas of EN ISO 13370. This standard assumes that heat flow across on and below grade floors and below grade walls varies harmonically:

$$\Phi = A\left[U_{red}\left(\theta_{i,m} - \theta_{e,m}\right) - L_{pi}\, \hat{\theta}_i \cos\left(2\pi \frac{m - \tau + \varphi_{\theta_i}}{12}\right) + L_{pe}\, \hat{\theta}_e \cos\left(2\pi \frac{m - \tau - \varphi_{\theta_e}}{12}\right)\right] \quad (5.29)$$

In that formula $\theta_{i,m}$ and $\theta_{e,m}$ are the annual mean in- and outside temperature, $\hat{\theta}_i$ and $\hat{\theta}_e$ the annual amplitudes, m the month considered (1 for January, etc). t is steady state time shift compared to a cosine, φ_{θ_e} and φ_{θ_i} the additional time shifts due to inertia and L_{pi} respectively L_{pe} the periodic heat transfer coefficients between the amplitudes of the in- and outside temperature and the heat flow amplitudes they induce. Time shifts φ_{θ_e} and φ_{θ_i} are given by:

$$\varphi_{\theta_e} = 1.5 - 0.42 \ln\left(\frac{\delta}{d_t + \delta}\right) \qquad \varphi_{\theta_i} = 1.5 - \frac{6}{\pi} \arctan\left(\frac{d_t}{d_t + \delta}\right) \quad (5.30)$$

with δ the annual harmonic temperature penetration depth in the soil, equal to:

$$\delta = \sqrt{3.15 \cdot 10^7\, \lambda_{gr} / \left[\pi (\rho c)_{gr}\right]}$$

The periodic heat transfer coefficients in turn are given by (for the symbols, refer to the reduction factors and the reduced thermal transmittance):

Floor on grade, no perimeter insulation

Temperature amplitude outside

$$L_{pe} = \frac{0.37\, P\, \lambda_{gr}}{A_{fl}} \ln\left(\frac{\delta}{d_t} + 1\right)$$

Temperature amplitude inside

$$L_{pi} = \frac{\lambda_{gr}}{d_t} \sqrt{\frac{2}{\left(1 + \delta/d_t\right)^2 + 1}}$$

Floor on grade, perimeter insulation

Temperature amplitude outside

Perimeter insulation horizontally (*D* width in m):

$$L_{pe} = \frac{0.37 \, P \, \lambda_{gr}}{A_{fl}} \left\{ \left[1 - \exp\left(-\frac{D}{\delta}\right)\right] \ln\left(\frac{\delta}{d_t + d'} + 1\right) + \exp\left(-\frac{D}{\delta}\right) \ln\left(\frac{\delta}{d_t} + 1\right) \right\}$$

Perimeter insulation vertically (*H* its height):

$$L_{pi} = \frac{0.37 \, P \, \lambda_{gr}}{A_{fl}} \left\{ \left[1 - \exp\left(-\frac{2H}{\delta}\right)\right] \ln\left(\frac{\delta}{d_t + d'} + 1\right) + \exp\left(-\frac{2H}{\delta}\right) \ln\left(\frac{\delta}{d_t} + 1\right) \right\}$$

Temperature amplitude inside

Same as floor on grade, no perimeter insulation

Floor above crawlspace

Temperature amplitude outside

$$L_{pe} = \frac{U_{o,floor}}{A_{fl}} \left(\frac{0.37 \, P \, \lambda_{gr} \, \ln(\delta/d_{gr} + 1) + U_x \, A_{fl}}{\lambda_{gr}/\delta + U_x + U_{o,floor}} \right)$$

Temperature amplitude inside

$$L_{pi} = \left(\frac{1}{U_{o,floor}} + \frac{1}{\lambda_{gr}/\delta + U_x} \right)^{-1}$$

Heated basement, living space walls and floors below grade

Temperature amplitude outside

$$L_{pe} = \frac{0.37 \, P \, \lambda_{gr}}{A_{fl}} \left\{ 2\left[1 - \exp\left(-\frac{H}{\delta}\right)\right] \ln\left(\frac{\delta}{d_{tw}} + 1\right) + \exp\left(-\frac{H}{\delta}\right) \ln\left(\frac{\delta}{d_t} + 1\right) \right\} H \, i$$

with *H* the height between grade and the underside of the basement floor or the floor below grade of a living space

Temperature amplitude inside

$$L_{pi} = \frac{\lambda_{gr}}{d_t} \sqrt{-\frac{2}{(1+\delta/d_t)^2 + 1}} + \frac{H \, P \, \lambda_{gr}}{A_{fl} \, d_{tw}} \sqrt{-\frac{2}{(1+\delta/d_{tw})^2 + 1}}$$

Floor above an unheated basement

Temperature amplitude outside

$$L_{pe} = U_{o,floor} \frac{0.37\, P\, \lambda_{gr}\left[2 - \exp(-H/\delta)\right] \ln\left(\delta/d_{gr} + 1\right) + (H_{bas} - H)\, P\, U_{w,bas} + 0.33\, n\, V}{(A_{fl} - H\, P)\, \lambda_{gr}/\delta + (H_{bas} - H)\, P\, U_{w,bas} + 0.33\, n\, V + A_{fl}\, U_{o,floor}}$$

with H the height between grade and the underside of the basement floor and H_{bas} the height of the basement between the underside of the basement floor and the underside of the floor above

Temperature amplitude inside

$$L_{pi} = \left\{\frac{1}{U_{o,floor}\, A_{fl}} + \frac{1}{(A + H\, P)\, \lambda_{gr}/\delta + (H_{bas} - H)\, P\, U_{w,bas} + 0.33\, n\, V}\right\}^{-1}$$

5.2.2.4 Moisture tolerance

Floors on grade, crawlspaces and basements face a series of moisture sources and potential moisture problems: groundwater, capillary water in the soil, seeping rain water in case of hardly permeating soil and/or ground surfaces levelling towards the building, facade wall runoff, 100% relative humidity in the soil, mould growth because of a too high relative humidity indoors, surface condensation and, interstitial condensation in case of thermally insulated floors on grade, floors above crawlspace or unheated basement and basement outer walls plus floors. Crawlspaces show additional problems such as evaporation from the bottom and water vapour inflow across the ground floor in residential spaces. Typical performance requirements are:

- Basements
  Depending on function.
  - No leakage of seeping rain, groundwater and rain run-off if basements are used as residential spaces or accommodate functions that demand dry conditions. Leakage limited to quantities that neither harm nor cause unacceptable damage if figuring as storage space and parking or housing other non-critical uses.
  - No unacceptable mould growth, surface and interstitial condensation when used as residential space or accommodating critical functions, less restrictions for mould and surface condensation when used as storage space, parking, etc.
- Crawlspaces
  - No mould nor unacceptable surface and interstitial condensation harming the floor above
  - Vapour inflow across the floor into the living spaces above neither giving monthly mean vapour pressure excesses neither passing the indoor climate class 3/4 threshold nor inducing mould growth there.
- Floors on grade
  - No wetting by seeping rain water
  - No mould nor unacceptable surface and/or interstitial condensation

Groundwater

Groundwater exerts water pressure against basement floors and outer walls (ΔP_w). The result is saturated water flow to the inside, per m² given by:

$$g_w = \frac{\Delta P_w}{\sum_{i=1}^{n} \frac{d_j}{k_{w,j}}} \qquad (5.31)$$

with d_j layer thickness in m and k_{wj} saturated water permeability in kg/(s·m·Pa). Without appropriate measures, pressure flow results in outflow at the basement side. Avoiding that demands a watertight layer somewhere in the assembly. With other words, water tightening is the only way out to stop groundwater leakage, see design and execution.

Capillary soil water

If a basement contacts a capillary wet soil, then the floor and outer walls will turn wet if their suction potential succeeds the soil's one. In clay and loam, the likelihood is low but it is much higher in sandy soils. In general, moistening is minimized by water tightening the floor and finishing all vertical surfaces contacting the soil with a render with very high suction potential. That render will turn wet though keeping the stony material underneath dry.

Seeping rainwater

In general, excavations extend beyond what is needed for the basement or crawlspace. Afterwards, the wedges between the outer walls and the intact ground are filled with debris and loose soil, creating a very permeable layer that way. Rain on surfaces levelling to the building can fill that wedge, building up a temporary water head against the basement or crawlspace outer walls with leakage as a probable result (Figure 5.20). Three methods allow avoiding such leakages:

1. Retard inflow

Slower inflow diminishes water heads in the wedges. In fact, equilibrium means inflow equals permeation into the intact ground. Less permeation means lower water heads. The lower they are, the lower the water pressure on the below grade outer walls and the lower the leakage probability. A way to retard inflow is by paving the wedge with flagstones that slope away from the building. That solution is most effective with well permeating intact grounds.

Figure 5.20. Seeping rainwater wedge around a basement.

2. Activating outflow

Drainage does that. A filtering layer is placed around the basement or crawlspace, which at foundation depth serves a system of sloped perforated pipes, coupled to the sewerage system. During rain, the pipes collect permeating water and conduct it to the sewers. See design and execution.

3. Tighten basement or crawlspace

Here, tightness against water heads is the objective. This demands watertight wall finishes or basements and crawlspaces constructed with materials so water impermeable that evaporation to the inside allows keeping the waterfront away from the inside. For that, water flow rate at interface x in the wall must equal vapour flow rate from x to the inside. For a single layered wall, that presumes the following relationship between water permeability and the vapour resistance factor:

$$k_w \frac{H}{x} = \frac{p_{sat,x} - p_i}{1/\beta_i + \mu N (d-x)} \quad \text{giving} \quad k_w \approx \frac{p_{sat,x} - p_i}{\left[1/(\beta_i\, x) + \mu N (d/x - 1)\right] H} \quad (5.32)$$

with H water head in m, d wall thickness in m, $p_{sat,x}$ water vapour saturation pressure at interface x and p_i vapour pressure in the basement or crawlspace. See design and execution.

Cavity and outside surface run-off

Cavity side and outer surface run-off along the facade walls can give infiltration in the basement or crawlspace. Without or with wrongly mounted tray below the cavity, cavity side run-off collects above the basement or crawlspace walls and may leak to the inside. The same thing will happen when outside surface run-off collects on protruding watertight layers just below the ground floor (Figure 5.21). The remedy is simple: care for correctly mounted cavity trays, avoid protruding watertight layers.

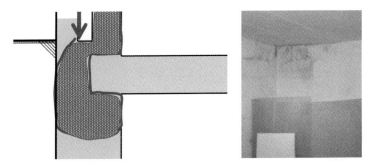

Figure 5.21. Leakage by missing cavity tray (left) or protruding waterproof barrier below the floor above (right).

Mould and surface condensation

Mould risk on walls and floors nears one when the monthly mean relative humidity at the surface passes 80%. Because of the high thermal inertia in crawlspaces and basements, this can become a real possibility during the warm months, on condition that relative humidity outdoors reaches high values, as is the case in the moderate climate of north western Europe.

Table 5.6. Semi-detached dwelling, ground floor 7.2 × 10.8 m², free perimeter 25.2 m. Basement relative humidity, climate of Uccle, Belgium, temperature in the living spaces 21 °C.

Month	Outside temperature	Outside RH	Relative humidity basement (%)		
			Walls dry		Walls wet
		Ventilation	0.1 ach	1 ach	1 ach
J	3,2	91	58	63	94
F	3,9	91	59	64	94
M	5,9	88	61	66	95
A	9,2	85	65	70	95
M	13,3	81	70	73	96
J	16,3	78	74	76	96
J	17,6	77	77	78	97
A	17,6	78	78	79	97
S	15,2	81	74	77	96
O	11,2	85	69	73	96
N	6,3	88	62	67	95
D	3,5	91	58	64	94

Consider a semi-detached dwelling with a 77.8 m² large basement having a free perimeter of 25.2 m, a 14 cm thick concrete floor and 29 cm thick masonry outer walls. The floor above is insulated with 3 cm PUR. In a first step, walls and floor are assumed dry. For a ventilation rate 0.1 and 1 ach, solving the combined vapour/heat balance gives the monthly mean relative humidity of Table 5.6. Even when moderately ventilated and dry, 80% is approached in summer. More ventilation is no solution, on the contrary, relative humidity increases! The reason is that the basement is cooler than outdoors. As the floor and lower parts of the walls in a basement are even colder than the air, both explain why basements not only feel clammy in summer but also why mould may develop on the floor and on stored goods such as paper. Wet basements are a true problem, see Table 5.6.

Interstitial condensation

Sometimes basements are used as residential space. In such case, insulating floor and walls has become a common practice. That may be done outside in contact with the soil, then called perimeter insulation, or inside.

Basement air-dry

Perimeter insulation

With perimeter insulation, interstitial condensation should not be a problem, though water from outside must be prevented from penetrating behind the insulation layer. What insulation material is to be used? Less permeable soils without drainage around the building demand boards (1) that do not suck water, (2) have a high diffusion resistance and (3) whose water-filled surface pores are not destroyed by frost. Extruded polystyrene (XPS) meets these requirements.

5.2 Performance evaluation

Cellular glass fails for (3). If the terrain is flat and the soil very permeable, then the probability of rain water seeping into the wedges around the basement is practically zero and materials such as highly water-repellent dense mineral wool boards can be used.

Inside insulation

Inhabited basements not only should offer good thermal comfort, usage also creates a vapour excess indoors, allowing insertion in one of the indoor climate classes. If the walls are then insulated at the inside, the contact temperature just below grade between the wall and the insulation will shift direction outside temperature, whereas deeper it will go direction annual mean temperature. At the soil side, relative humidity equals 100%. What value will be noted between a 'dry' basement wall and the inside insulation depends on the indoor climate class and the solution applied.

As Figure 5.22 shows for the moderate climate of Uccle, for a timber frame, the bays filled with mineral wool and finished with gypsum board, climate class 2 gives up to 100% relative humidity between 0 and 0.5 m below grade at the backside of the masonry with moisture deposited from December to March. Indoor climate class 3 maintains 100% relative humidity from November to June with moisture deposited from November to March. The two are not tolerable. To exclude mould and timber rot, monthly mean relative humidity at the wall/insulation interface should stay below 85%. This, however, is not doable, even not with a perfect vapour retarder between insulation and gypsum board. If such vapour retarder is anyhow applied, relative humidity in the insulation package touches 97% in summer. Clearly, not all solutions function properly. Those that do are listed in Table 5.7.

Basement capillary wet

Perimeter insulation

No problem, as perimeter insulation is mounted during construction, which allows water proofing all basement walls properly.

Inside insulation

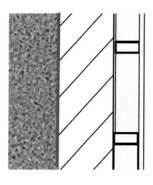
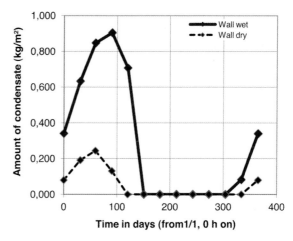

Figure 5.22. Basement wall insulated inside, timber frame, the bays filled with 6 cm thick glass fibre bats and finished with gypsum board, indoor climate class 2, interstitial moisture deposit.

With inside insulation, except XPS and cellular glass, basement walls have to be vapour-tightened first, see Table 5.7. If not, summer relative humidity may reach very high values, not only in the insulation but also at the backside of the inside lining and in the basement, see Figure 5.22. Moreover, the insulation must be mounted correctly (see the chapter on massive walls).

Table 5.7. Basement walls insulated inside, moderate climate: vapour retarder quality.

Solution	Vapour retarding quality		
	Indoor climate class 1	Indoor climate class 2	Indoor climate class 3
Basement walls correctly water-tightened at the outside, thus: air-dry			
Timber frame, bays filled with mineral wool, gypsum board finish	None	Do not apply	
Mineral wool/gypsum board composite glued against the wall	None	E1 $(\mu d)_{eq} \leq 5$ m	E2 $5 < (\mu d)_{eq} \leq 25$ m
EPS, PUR glued against the wall and finished with gypsum board or sprayed gypsum	None	None	E1
XPS glued against the wall and finished with gypsum board or sprayed gypsum	None	None	None
Cellular glass, glued against the wall and finished with gypsum board or sprayed gypsum	None	None	None
Basement wall capillary wet (water-tighted neither at the outside nor at the inside)			
Timber frame filled with mineral wool, gypsum board finish		Do not apply	
Mineral wool/gypsum board composite glued against the wall after vapour-tightening the basement wall at its inside	None	E1	E2
EPS, PUR, glued against the wall and finished with gypsum board or sprayed gypsum after vapour-tightening the basement wall at its inside	None	None	E1
XPS glued against the wall and finished with gypsum board or sprayed gypsum	None	None	None
Cellular glass, glued against the wall and finished with gypsum board or sprayed gypsum	None	None	None

5.2 Performance evaluation

Vapour from the crawlspace entering the living space

Crawlspace air always carries water vapour. Conservation of mass gives the following relations between incoming and outgoing air:

For the groundfloor

$$G_{aei} + G_{aie} + G_{aic} = 0$$

$$\text{or} \quad -(K_{aei} + K_{aie} + K_{aic})P_{ai} + K_{aic}P_{ac} = -K_{aei}P_{aei} - K_{aie}P_{aie} - K_{aic}p_{Ti}$$

with:

$$K_{aei} = \frac{a_{ei}}{\left[\text{abs}(P_{aei} - P_{ai})\right]^{0.5}}; \quad K_{aie} = \frac{a_{ie}}{\left[\text{abs}(P_{aie} - P_{ai})\right]^{0.5}};$$

$$K_{aic} = \frac{a_{ic}}{\left[\text{abs}(P_{ac} + p_{Ti} - P_{ai})\right]^{0.5}}; \quad p_{Ti} = 9.81 \frac{100\,000}{287} \frac{2}{T_i}$$

In the crawlspace

$$G_{aec} + G_{ace} + G_{aic} = 0$$

$$\text{or} \quad K_{aic}P_{ai} - (K_{aec} + K_{ace} + K_{aic})P_{ac}$$
$$= -K_{aec}(P_{aec} + p_{Te} - p_{Ti}) - K_{ace}(P_{ace} + p_{Te} - p_{Ti}) + K_{aic}p_{Ti}$$

with:

$$K_{aec} = \frac{a_{ec}}{\left[\text{abs}(P_{aec} + p_{Te} - P_{ac} - p_{Ti})\right]^{0.5}}; \quad K_{ace} = \frac{a_{ce}}{\left[\text{abs}(P_{ace} + p_{Te} - P_{ac} - p_{Ti})\right]^{0.5}};$$

$$K_{aic} = \frac{a_{ic}}{\left[\text{abs}(P_{ai} - P_{ac} - p_{Ti})\right]^{0.5}}; \quad p_{Ti} = 9.81 \frac{100\,000}{287} \frac{2}{T_e}$$

The steady state water vapour balances look like:

In the living space

$$\frac{0.62 \cdot 10^{-5}}{A_{fl}}\left[G_{aei}p_e + G_{aci}p_c - (G_{aei} + G_{aci})p_i\right] + \frac{p_c - p_i}{Z_{fl}} + \frac{G_{vP}}{A_{fl}} = 0$$

In the crawlspace

$$\frac{0.62 \cdot 10^{-5} G_{aec}}{A_{fl}}(p_e - p_c) + \frac{p_b - p_c}{Z_{b,c}} + \frac{p_i - p_c}{Z_{fl}} = 0$$

In these equations, G_{aec} is the outside air entering the crawlspace, G_{ace} the air leaving the crawlspace to outside, G_{aci} the air intruding in the living space from the crawlspace and G_{aie} the air flowing from the living space to outside, all in kg/s. K_{aec}, K_{ace}, K_{aic}, K_{aei} and K_{aie} are related air permeances and P_{aec}, P_{ace}, P_{ac}, P_{ai}, P_{aei} and P_{aie} air pressures causing the air flows.

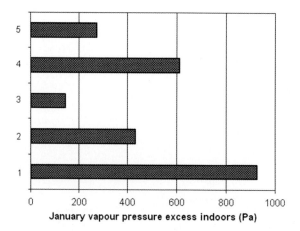

Figure 5.23. Limiting vapour inflow from a crawlspace into the ground floor, vapour excess indoors showing the effectiveness of different measures (1 = no measures, 2 = floor above air-tightened, 3 = ground cover in the crawlspace, 4 = floor above insulated, 5 = warm crawlspace).

a_{xy}-s represents the air permeance coefficients and, p_{Te} and p_{Ti} the stack pressures at the vents in the crawlspace and trickle vents in the living space above. Height between both is assumed 2 m. p_e, p_c and p_i are vapour pressure outdoors, respectively vapour pressure in the crawlspace and vapour pressure in the living space, all in Pa. p_b is the vapour pressure below the soil cover in the crawlspace. If that soil is capillary wet, then p_b equals saturation pressure $p_{sat,b}$ at soil temperature. Z_b stands for the diffusion resistance of the soil cover and Z_{fl} for the diffusion resistance of the floor above the crawlspace. G_{vP} is the vapour released in the living space above in kg/s. In case the relative humidity in the crawlspace is the variable looked for, then also the thermal balance must be solved.

The mean vapour balance underscores that water vapour inflow into the living space above intensifies with higher air inflow from the crawlspace, resulting in more vapour pressure excess indoors, as Figure 5.23 shows for January in a moderate climate. Of all possible measures to avoid inflow, a vapour retarding soil cover in the crawlspace looks most effective. Second in line is a warm crawlspace or air tightening the floor above. Least effective is floor insulation, a measure that lowers temperature and increases relative humidity in the crawlspace.

5.2.2.5 Thermal bridging

Because the load bearing capacity required conflicts with continuity of the insulation layers, intersections between foundations, below grade walls, facade walls and floors typically figure as preferential thermal bridges. At the facade, the problem is mitigated by (1) extending the insulation layer in the outer walls below grade, (2) finishing the foundation walls with perimeter insulation, (3) applying manufactured thermal cut solutions in the outer walls (Figure 5.24). As Table 5.8 shows, (3) is most effective (lowest linear thermal transmittance, highest temperature factor). However, restricted load bearing capacity of the insulation material, cellular glass, limits application to moderately loaded walls. If necessary, a perimeter beam above the break should distribute loads equally (Figure 5.25).

5.2 Performance evaluation

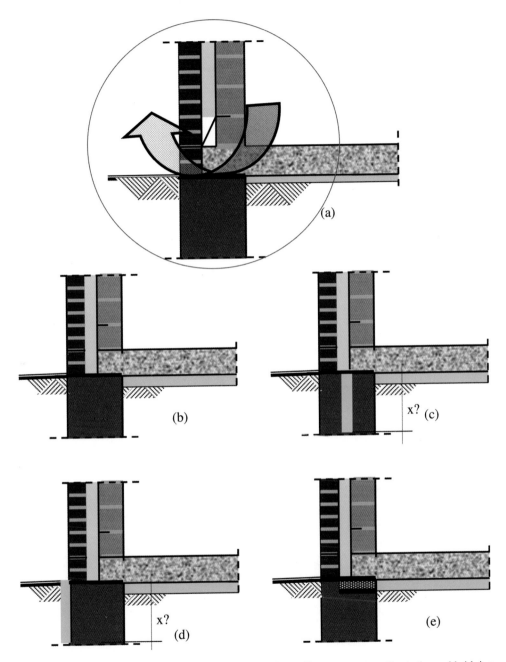

Figure 5.24. Node foundation wall/floor on grade/cavity wall: measures to mitigate thermal bridging (also see Table 5.8).

Table 5.8. Thermal bridging in the node 'foundation wall/facade wall/floor' (Figure 5.24).

Node detail	Thermal bridging		
	Heat loss $\theta_e = 0\,°C$, $\theta_i = 20\,°C$ W/m	ψ-value W/(m·K)	Temp. Factor (corner coldest spot)
Reference: heat loss according to ISO EN 13370 for a floor with $L \times W = 5.39 \times 1\,m^2$, in combination with a cavity wall (1.08 m considered in the calculation)	31.1	–	–
Bad workmanship ((a) in Figure 5.24). Floor contacting the veneer wall, cavity fill not touching the floor	37.5	0.33	0.74
Correct workmanship ((b) in Figure 5.24). Cavity extending to the floor's underside, cavity fill starting there	30.2	–0.05	0.85
Cavity fill extending below grade till foundation footing ((c) in Figure 5.24)	28.4	–0.14	0.88
Correct workmanship, perimeter insulation ((d) in Figure 5.24)	29.7	–0.07	0.86
Thermal cut in the inside leaf, joining the floor insulation ((e) in Figure 5.24)	27.2	–0.19	0.91

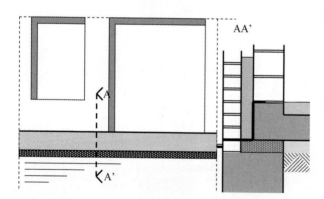

Figure 5.25. Perimeter beam above a thermal break.

5.2.3 Building physics: acoustics

In case the basement serves as parking space, sound transmission loss for airborne noise in the 500 terts band ($R_{w,500}$) for the floor above should equal or exceed 56 dB(A). A thick heavyweight concrete deck will guarantee that level of performance.

5.2.4 Durability

Thermally few things happen. Changes in outside temperature are damped to such extend by the soil that fluctuations left are too small to induce cracking in below grade parts Also hygrically there is hardly anything to worry about, at least as long as the below grade parts keep a more or less constant moisture content. Fresh concrete of course suffers from chemical shrinkage, which should be neutralized by an adapted casting scheme. However, if for any reason air-dry walls in basements become wet, then swelling may cause irreversible damage: cracks in the inside render, plaster profiles at corners pushed away, peeling paint, etc.

5.2.5 Fire safety

In medium and high rises, the floor at grade separates two fire compartments: the basement and the spaces above. That demands a floor system and load bearing structure in the basement with overall fire resistance of 90′ or more.

If successive basement floors serve as parking lots, then staircases and lifts may not give direct connection to the upper floors. The only solution allowed is access to the ground floor and change over there to another staircase and other lifts. In the basements and at the ground floor, staircases and lifts must additionally start and end in a lock, whose walls and doors guarantee a fire resistance of at least 30′. One must also install a ventilation system in the parking lots that cares for effective smoke and heat removal in case of fire. And, finally, for fire loads touching high values, fire alarms and sprinklers have to be installed. Heating and cooling plants located in the basement are subjected to even more severe fire safety requirements.

5.2.6 Soil gases

In regions with primary rock soils, keeping radon out of the basements could be done in theory by a perfectly tight basement, crawlspace or floor on grade slab construction. In practice, achieving complete tightness is close to impossible, which is why an alternative solution is more appropriate: combining a well-tightened slab with gravel drainage below, coupled to a sub-floor ventilation system that depressurizes the gravel compared to indoors and prohibits radon inflow that way.

5.3 Design and execution

5.3.1 Basements

Low rises have cast on site concrete basements, precast concrete basements or basements combining concrete floors with 30 cm thick massive concrete block walls. Massive blocks have the advantage compared to hollow blocks that they do not offer storage volume for leaking water. Water conductive massive blocks in gravel concrete however should not be used as the smallest crack in the watertight or water repelling exterior render will end in water leakage.

We discuss waterproofing of basements below. In any case, to avoid rising damp in the walls above grade or leakage from rain run-off into the basement, all basement walls receive a waterproof barrier just below the ground floor. When the exterior facade finish is not capillary,

Figure 5.26. Outside protruding waterproof barrier below the floor at grade causes rain leakage into the basement.

that barrier however should not protrude to the outside. In fact, a leakage path just above the barrier into the basement may be created that way with the small run-off water head on top of the protruding barrier as driving force (Figure 5.26).

In countries with a timber tradition, basements are sometimes constructed using timber framed walls. Foundations in such case consist of a gravel bed with concrete floor above, upon which the timber-framed walls are mounted. The timber gets a protective treatment beforehand, while all below grade parts are water-tightened at the outside. Below grade timber constructions anyhow are strongly discouraged in termite-infested regions, starting south of Paris in Europe, and south of the 48th latitude in the USA. There, concrete is the best choice for basement construction and insulating the basement walls must be done at the inside, beyond termite reach. The timber-framed structure then starts some 5 cm above grade with between concrete and timber steel plate inserts as anti-termite-screen.

In medium and high rises, basements are constructed as stiff boxes in reinforced concrete, often with the deep or retaining walls that stabilize the excavation used as outer walls. The boxes allow redistribution over a large area of local loads exerted by the concrete or steel frame above. Usage of so-called watertight concrete should limit future water leakage (see drainage and water tightening).

In all cases, the earth conductor of the electrical installation must be installed below the foundation footings, foundation slabs or stiffening beams.

5.3.2 Drainages

5.3.2.1 In general

Drains for transporting seeping water include: (1) a watertight layer against the basement walls, (2) a vertical drain all around enclosed by filter fabric, (3) a subsystem of porous or perforated pipes, served by the drain and ending in the sewage system or a pump well (Figure 5.27). In addition, the ground around the building should slope away.

5.3 Design and execution

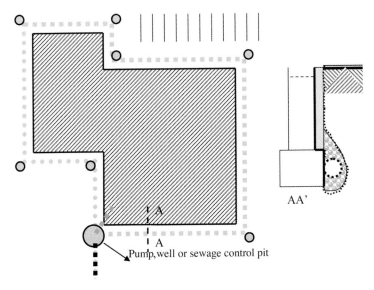

Figure 5.27. Drain.

5.3.2.2 Properties

Distinction must be made between the flow across the filter fabric and the flow in the backfill. Flow across the fabric depends on its water permeance: $P_f = k_{wf}/d_f$, with k_{wf} water conductivity and d_f thickness of the fabric. The backfill instead is characterized by a transmittance T_d, equal to $k_w d_d$, and representing the maximal flow across a 1 m deep backfill per meter run under a 1 m water head. k_{wd} is the water conductivity of the backfill and d_d its width.

5.3.2.3 Design

Filter fabric

The fabric must be water permeable but soil proof. Selection criteria used are:

$$\Psi_{95} \leq \alpha_1 \alpha_2 \alpha_3 \alpha_4 d_{85} \text{ (ground)} \qquad k_{wf} = \left(10^3 \ldots 10^5\right) k_{eg}$$

with k_{eg} water conductivity of the soil. Ψ_{95} means 95% of the fabric pores should be smaller than the value given by the right side of that inequality. d_{85} characterizes the sieve mesh where 85% of the soil grains touching the fabric pass through. The factor α_1 depends on the grain distribution: a value of 1 when covering a large interval ($d_{60}/d_{10} > 10$), a value of 0.8 in case of a very homogeneous soil ($d_{60}/d_{10} \approx 1$). α_2 characterizes soil compactness: 0.8 if densely packed, 1.25 if loosely packed. α_3 involves the water pressure gradient of the water table at the fabric in m water head: 1 for $\text{grad}(P_w) < 5$ m/m, 0.8 for $\text{grad}(P_w) = 5$ à 20 m/m, 0.6 for $\text{grad}(P_w) > 20$ m/m. α_4 clarifies the filter fabric function. If only active as filter, α_4 equals 1. If active as filter and drain, α_4 equals 0.3. Of course, all that presumes the soil's grain distribution is measured beforehand, which is not normal practice. Many drainage systems consequently do not function properly.

Backfill

The backfill must transport the maximum seeping flow expected at the basement walls (G_r) without sustained water head: $T_d \geq f G_f$, with f a safety factor (f.e.: $f = 1.5$). The difficulty with that equation is guessing the maximum seeping flow. For several weather stations, the maximum precipitation per m² noted over a period of for example 25 years is known. Translating that number into flows a backfill will have to accommodate however is not so simple due the distribution between surface flow and direct infiltration intervening. In sandy soil, direct infiltration is large; in loam and clay instead it hardly exists. Backfills consist of gravel, porous high-pressure polyethylene boards, expanded polystyrene boards or geo-composites.

Pipes

As said, the system of pipes collects and transports the drained water to the sewage system or a pump well (Fig. 5.27). Perforated flexible synthetic pipes are used. They get a slope of 0.5 to 1%, starting at the upper edge of the foundations. At all corners and where coupled to the sewage system, control pits are included. The system is dimensioned as a sewage pipe: section half full, but with continuous supply. Pipe diameter: 100 to 200 mm.

Waterproofing

To bring some system in the combination of drains and waterproof finishes, a division in five classes is forwarded together with an array containing the criteria, which allow selection of the right class for any application, see Tables 5.9 and 5.10.

Table 5.9. The five classes of moisture protecting measures for basements with depths to 3 m.

Class	Waterproofing	Backfill
1.	Waterproof encasement	Porous plastic foam boards, $T_d > 1$ l/(m·s), pressure strength adapted to soil pressure present, filter fabric adapted to soil type
2.	Waterproof encasement Reinforced bitumen paste	Porous plastic foam boards, $T_d > 0.5$ l/(m·s), pressure strength adapted to soil pressure present, filter fabric adapted to soil type
3.	Synthetic resin Moisture repellent mortar Mortar layer finished with bituminous emulsion	Porous plastic foam boards, $T_d > 0.5$ l/(m·s), pressure strength adapted to soil pressure present, filter fabric adapted to soil type
4.	Reinforced bitumen paste, protected by perimeter insulation Synthetic resin Moisture repellent mortar Mortar layer finished with bituminous emulsion	None
5.	None	Porous plastic foam boards, $T_d > 0.1$ l/(m·s), pressure strength adapted to soil pressure present, filter fabric adapted to soil type

5.3 Design and execution

Table 5.10. Selection criteria.

Soil	Slope and surface permeance of the ground		
	Impermeable, sloping towards the building	Flat ground	Sloping away from the building
Basement function not allowing wall wetness			
Very permeable	Class 2	Class 3	Class 4
Moderately permeable	Class 1	Class 2	Class 3
Less permeable	Class 1	Class 1	Class 2–3
Temporarily moist walls not a problem			
Very permeable	Class 3	Class 4	Class 4–5
Moderately permeable	Class 2	Class 3	Class 4–5
Lesspermeable	Class 2	Class 3	Class 3

Variables are basement function, slope and surface permeance of the ground surrounding the building and soil water conductivity. Class 1 stands for absolutely watertight. Class 5 assumes such high soil infiltration that basement tightening is not needed on condition the basement walls are brick-laid in massive blocks. For a more detailed description of all materials and all possible tightening solutions applied, reference is made to the literature and manufacturers information.

5.3.3 Waterproof encasement

Drains are applicable as long as seeping rain is the only water source involved. Instead, for basements below the water table, waterproof encasements of class 1 are needed. Possibilities offered are inside and outside encasements or the use of waterproof concrete.

5.3.3.1 Inside

With an inside encasement, waterproofing is added inside the below grade construction. A possible advantage is easier repair. A disadvantage is the outside walls and lowest floor stay wet, so organic acids and salts in the groundwater may attack the masonry or concrete. Two systems apply: (1) watertight membrane, protected by an inside reinforced concrete wall/floor structure, (2) waterproof rendering.

Waterproof membrane with inside reinforced concrete wall/floor support

Building is done in an excavation with well point dewatering. Once the load bearing structure is ready, the basement is water proofed at the inside using synthetic or SBS polymer bitumen membranes. An important membrane property should be good deformability to follow hygric deformation of the structure and to absorb differential settlement. The number of SBS polymer bitumen layers depends on the water head:

Water head (h_w) m	Number of layers
$h_w \leq 4$	2
$4 < h_w \leq 9$	3
$h_w > 9$	4

For synthetic membranes, the water head defines what thickness to use. With PVC:

Water head (h_w) m	Thickness mm
$h_w \leq 9$	1.5
$h_w > 9$	2.0

Once bonded, the waterproof membrane gets a protective layer, for example XPS-boards, after which the inside reinforced concrete support is cast. That support has to withstand water pressure. Upside floor pressure is transferred to the building by anchoring the concrete support in the basement walls some 30 cm above the highest ground water level measured over the last 25 or more years. The waterproof membrane has to join the horizontal waterproof barrier there. If the last is absent, problems with rising damp may surface after the building is finished. Once the support structure is ready, all non-bearing basement partitions and staircases are built. A critical point with inside waterproof encasements is the membrane overlap in the edges and corners. Also see Figure 5.28.

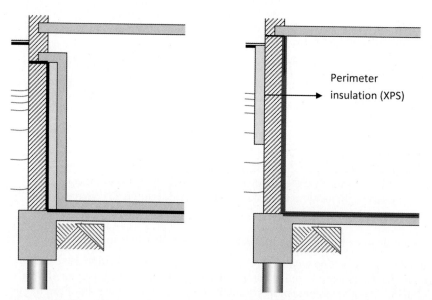

Figure 5.28. Inside encasements: left waterproof membrane with reinforced concrete support, right waterproof render.

5.3 Design and execution

Waterproof render

The use of a waterproof rendering at the inside is an alternative for small water heads (≤ 2 m). The render typically is a sand/cement/filler/synthetic resin mixture with low open porosity, very fine pores and low creep. That way water conductivity is kept very low. Once bonded, the render should remain deformable enough as to absorb limited hygric deformation and limited settlement of the substrate without cracking. This is critical, which is why waterproof renders are only used in cases where expected remaining deformation and settlement is small as is the case in existing buildings suffering from ground water leakage. Render bonding to the substrate should be of such strength that water pressure is withstood. See Figure 5.28.

Although not advisable, when used in new construction, rendering is done first. All non-bearing partitions are then brick-laid. This must be done without perforating the render. Fix doorframes for example at no point through the render into the walls and do not fix doorsteps in the floor. This is critical. When applied in existing building, all inside partitions must also be rendered.

Waterproof rendering is a less than safe solution. It is not to be advised in new construction.

5.3.3.2 Outside

Here, waterproofing is realized outside the below grade construction. The main advantages of an outside encasement are that the construction remains dry and that the building directly bears the water pressure and upside force. Repair work however is hardly possible afterwards, which is why good workmanship is an absolute prerequisite.

Building is done again in an excavation with well point dewatering. Above the foundations comes a slab, which protrudes out to out some 30 cm at all sides of the building floor area. Once cast, the slab is carefully flattened. After binding, a horizontal membrane of run-resistant synthetic felt or SBS polymer bitumen is adhered to the concrete and a protective layer added. Then the properly reinforced concrete foundation floor is cast, dimensioned for the pressure on the membrane nowhere to exceed 1 MPa. As soon as the below grade outer walls are at safe height compared to the highest ground water level measured over for example the past 25 years, their outside is also water proofed using synthetic felt or SBS polymer bitumen membranes. For the number of SBS polymer bitumen layers or synthetic PVC felt thickness the same rules as for inside encasements apply. The vertical membranes get as protection a thermal insulation layer, finished by masonry walls. Most critical in execution is the overlap between horizontal and vertical membranes. Figure 5.29 shows a solution.

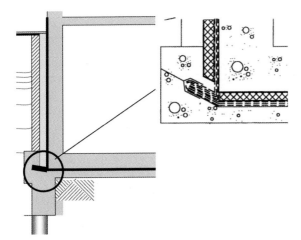

Figure 5.29. Outside encasement.

5.3.4 Waterproof concrete

Outside encasements are expensive, which is why building below grade using water impervious materials such as waterproof concrete is an attractive alternative. Such concrete is not really water impervious. The water conductivity is as low that drying to the inside offsets moistening from outside.

Waterproof concrete demands precaution. Thickness of the walls must be large enough, with 30 cm as a minimum. Whether more is needed, depends on the mechanical strength, the stiffness demanded and water resistance needed. Aggregates must be carefully chosen and mixing permanently controlled (correct granulometry, addition of fillers, etc.). During casting, successive layers have to be vibrated to get optimal density. Once casted the concrete is kept wet the first 7 days to limit shrinkage. Shrinkage cracks in fact kill water tightness. Casting joints are tightened using preformed synthetic profiles. That way, the route the water has to follow across the concrete is extended. The result is a higher water resistance and an easier balance between wetting from the outside and drying to the inside. Additional injection hoses nevertheless are inserted around all casting joints, allowing tightening remaining leaks by synthetic resin injection after well point dewatering stops. Settling joints demand analogous precautions. They have to be planned at regular intervals and tightened using the same preformed profiles with injection hoses at both sides as for the casting joints (Figure 5.30).

For pipe passages to be watertight, casings with flanges and clamping closure are used (Figure 5.31). Actually, the same casings are used for pipes passing inside and outside encasements.

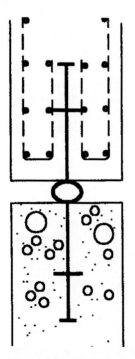

Figure 5.30. Settling joint.

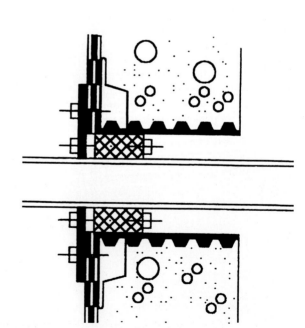

Figure 5.31. Pipe passage through water-proof concrete.

5.4 References and literature

[5.1] AJ Information Library, *The Art of Construction, phase 1, Substructure, AJ 11.* March 1981.

[5.2] WTCB, TV 147, *Funderingen van huizen.* Juni 1983 (in Dutch).

[5.3] Standaert, P. (1984). *Funderingskoudebruggen, Rapport in het kader van de samenstelling van een catalogus "Bouwkundige details".* Laboratorium Bouwfysica, 4 p. (in Dutch).

[5.4] *Assessment of the energy savings potential of building foundation research.* ORNL/Sub/85-27497/1, 1986, 41 p.

[5.5] *Building Foundation Design Handbook.* ORNL/Sub/86-72143/1, 1988, 349 p.

[5.6] Lutz, Jenisch, Klopfer, Freymuth, Krampf (1989). *Lehrbuch der Bauphysik.* B. G. Teubner, Stuttgart, 711 p. (in German).

[5.7] Cziesielski, E. (1990). *Lehrbuch der Hochbaukonstruktionen.* B. G. Teubner, Stuttgart, 627 p. (in German).

[5.8] Frick, Knöll, Neumann, Weinbrenner (1990). *Baukonstruktionslehre, Teil 1.* B. G. Teubner, Stuttgart, 588 p. (in German).

[5.9] WTCB, TV 190, *Bescherming van ondergrondse constructies tegen infiltratie van oppervlaktewater.* December 1993, 43 p. (in Dutch).

[5.10] Herrmann, M. (1995). *Wärmeverluste erdreichberührter Bauteile.* Diplomarbeit EMPA (in German).

[5.11] *IEA-Annex 24, Final Report, Vol. 1, Task 1: Modelling, Addendum.* ACCO, Leuven, 1996.

[5.12] Krarti, M., Claridge, E., Kreider, J. (1995). *Frequency response analysis of ground coupled building envelope surfaces.* ASHRAE Transactions, Vol. 101, part 1, p. 355–369.

[5.13] Krarti, M., Sangho, C. (1996). *Simplified method for Foundation Heat Loss Calculation.* ASHRAE Transactions, Vol. 102, part 1.

[5.14] Building Science Corporation (1997). *Builder's Guide, Mixed Climates.*

[5.15] Adnan, A. Al-Anzi, Krarti, M. (1997). *Evaluation of the Thermal Bridging Effects on the Thermal Performance of Slab-on-Grade Floor Foundation.* ASHRAE Transactions, Vol. 103, part 1.

[5.16] Building Science Corporation (1998). *Builder's Guide, Hot-Dry & Mixed-Dry Climates.*

[5.17] Hens, H. (1998). *Toegepaste Bouwfysica 2, bouwdelen.* ACCO, Leuven, 341 pp. (in Dutch).

[5.18] NBN EN ISO 13370 (1998). *Warmte-eigenschappen van gebouwen. Warmte-uitwisseling via de grond. Berekeningsmethoden.* Brussels, BIN (in Dutch).

[5.19] Building Science Corporation (2000). *Builder's Guide, Hot-Humid Climates.*

[5.20] ASHRAE (2001). *Proceedings of the Performance of Exterior Envelopes of Whole Buildings VIII Conference.* Florida (CD-ROM).

[5.21] Janssen, H. (2002). *The influence of soil moisture transfer on building heat loss via the ground.* Doctoral Thesis, K. U. Leuven.

[5.22] Rantala, J., Leivo, V. (2003). *Moisture behavior of slab-on ground structures, Research in Building Physics* (Eds.: J. Carmeliet, H. Hens, G. Vermeir). Balkema Publishers (p. 345–351).

[5.23] Hens, H. (2004). *Toegepaste Bouwfysica 1, randvoorwaarden, prestaties, materiaaleigenschappen.* ACCO, Leuven, 186 pp. (in Dutch).

[5.24] ASHRAE (2004). *Proceedings of the Performance of Exterior Envelopes of Whole Buildings IX Conference.* Florida (CD-ROM).

[5.25] Boardman, C., Glass, S., Charles, C. (2010). *Estimating Foundation Water Vapor Release Using a Simple Moisture Balance and AIM-2: Case Study of a Contemporary Wood-Frame House, Proceedings of the Performance of Exterior Envelopes of Whole Building XI Conference.* Florida (CD-ROM).

[5.26] Hens, H. (2010). *Applied Building Physics. Boundary conditions, Building Performance and Material Properties.* Ernst & Sohn (A Wiley Company), Berlin.

[5.27] Hens, H. (2010). *Prestatie-analyse van bouwdelen 1.* Uitgeverij ACCO, 327 pp. (in Dutch).

6 Structural options

6.1 In general

The term 'structural' indicates the building's load bearing elements. A structural system is designed to carry the horizontal and vertical loads and transfer them to the foundations. In high-rises it defines much of the design. Horizontal forces in fact are of such importance that limited freedom is left in choosing a system and designing the building's shape.

In general two extremes are at the designer's disposal: (1) spreading the loads over as much vertical parts as possible, (2) concentrating the loads. The first results in massive structures, the second in skeleton structures. Both have advantages and disadvantages:

- Skeleton structures
    + Create great freedom in floor lay-out. They are therefore preferred when functional requirements include easy future floor lay-out reorganisation
    − Make horizontal stability more difficult. Constructing a skeleton as a series of frames gives some relief although limited column dimensions persists in restricting stiffness (Figure 6.1).
- Massive structures
    − Results in rigid floors lay-outs. A well-balanced choice of load-bearing walls may give some flexibility but never the freedom skeleton structures offer. Massive construction should therefore only be recommended when functional requirements make future floor lay-out reorganisations unlikely.
    + Make horizontal load bearing capacity and stiffness easily realisable

The structural system chosen impacts choice of material. For skeleton constructions, steel is an option, not so for massive structures. Reinforced concrete and pre-stressed concrete instead are applicable for both, whereas masonry presumes massive structures. Timber also can be used for both.

In reality the choice between both structural types is not the one or the other. The two are often mixed, which allows combining the advantages and avoiding the disadvantages.

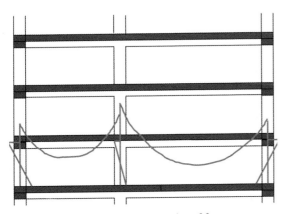

Figure 6.1. Skeleton acting as a series of frames.

6.2 Performance evaluation

6.2.1 Structural integrity

Structural systems are designed to guarantee building stability, safety against rupture and collapse, acceptable deformation under the weight carried and tolerable acceleration when dynamically excited. Design loads include:

In vertical direction

1. *Own weight*
   The weight of the structural system (slabs, beams, columns, walls)
2. *Death load*
   The weight of non-loadbearing walls and all finishes. Floor finishes include ceiling, screed and cover. For load and non-loadbearing walls plaster layers or any other wall finish may be used. Also non load- bearing facades are a dead load.
3. *Snow load*
   This is important for roofs. The snow load to consider largely depends on the climate zone. The greatest snow height recorded over the past 50 years, multiplied with a safety factor 1.5, is in general taken as design reference.
4. *Live load*
   Includes all function related loads a structural system has to carry once the building is in use. In residential buildings, useful load includes furniture, furnishing, people and the dynamic loads caused by stepping and running. In office buildings dead load also encompasses movable walls. For each building type, standards impose a value in N/m². For a building of several floors, useful load is reduced above a given number of storeys. The likelihood all floors are fully loaded in fact is close to zero. For design values, see the EN standards in Europe and their adaption to the specific situation in each country and, the International Codes in the USA.

In horizontal direction

1. *Obliqueness*
   Results from construction inaccuracies. Obliqueness causes parasite forces, comparable to fictitious horizontal loading (Figure 6.2):

$$P_H = P_V \, \mathrm{tg}(\phi) \approx P_V \frac{s}{H} \tag{6.1}$$

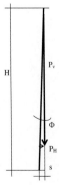

Figure 6.2. Obliqueness.

6.2 Performance evaluation

In both directions

1. *Wind load*
   Wind is a dynamic load with a broad spectrum, changing direction continuously and sucking on horizontal and moderately sloped surfaces. For steeper sloped surfaces suction turns into a compressive load at the windward side. Wind also exerts pressure on windward looking vertical and steeper sloped surfaces. On leeward looking vertical and steeper sloped surfaces and on surfaces parallel to the wind, pressure becomes suction.
   Any structural system has to be wind-stiff along its two main inertia axes. Design for wind load is based on maximum and extreme wind. At maximum load the building must maintain full fitness. Extreme winds may disrupt fitness however without causing rupture or collapse.

2. *Earthquakes*
   Earthquakes subject buildings to alternating dynamic horizontal and vertical displacements, complemented with a torsion inducing rotation. The most important requirements to be considered are stability and safety against rupture and collapse.

6.2.2 Fire safety

Fire counts as an exceptional thermal load. In low and medium rises, fire may not cause structural collapse within the time span needed for evacuation. Once evacuated, safety against collapse dropping below 1 is accepted if not incurring unwanted consequences. In high rises and buildings where function does not allow structural inadequacy, fire may not cause collapse and should not cause deformations such that the whole structure or parts of it become unusable.

Steel and concrete skeleton structures collapse when more plastic hinges are formed under fire than allowed by statics or when columns buckle. Plastic hinges appear when beams or columns heat up so strongly that the accompanying decrease in mechanical strength and increase in deformability turns them locally into plastic materials (Figure 6.3). For steel sections, plastic hinge formation starts when:

$$n_P = \frac{F\,\sigma_{E,20}/f_F}{\sigma_{max}} = 1 \tag{6.2}$$

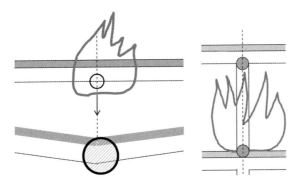

Figure 6.3. Concrete or steel skeleton: plastic hinge formation by fire.

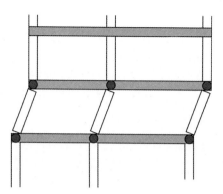

Figure 6.4. Collapse of a skeleton as a consequence of plastic hinge formation in the columns.

In that ratio $\sigma_{E,20}$ is the yield point at 20 °C, σ_{max} is the maximum stress in the section, F the form factor of that section and f_F becomes the factor representing the decrease in yield point with increasing temperature compared to the yield point value at 20 °C.

Plastic hinge formation makes fire safety to some extent predictable. Safety increases the more a structure is hyper-static. Columns are more critical than beams. They can buckle while two hinges turn them into pendulums. If in a skeleton without stiffening walls all columns develop two plastic hinges, collapse is unavoidable (Figure 6.4).

For low and medium rises, a fire resistance of 90′ is the objective. This means:

- Concrete
  - Structural system hyper-static
  - Concrete covering of main reinforcement in beams and columns of at least 55 mm
  - Concrete covering of main reinforcement in slabs at least 35 mm
  - Slabs more than 10 cm thick
- Steel
  - Structural system hyper-static
  - Columns and beams protected by non-burnable board material. 90′ demands a board thickness of 25 mm or more
- Timber
  - Structural system hyper-static
  - Timber structure massive or protected by non-burnable board material

For high rises and buildings with special functions, the whole structural system has to be built from non-combustible materials or should be protected so that all parts have a fire resistance far above 90′.

6.3 Structural system design

6.3.1 Vertical loads

In low and medium rises vertical loads cause few problems. For limited building heights, load bearing walls thickness in massive structures is defined by sound insulation and fire resistance rather than by strength and stiffness. Also in skeleton structures, where columns only have a load bearing function, a large freedom in sizing exists. In many cases, sections are based on modular coordination, uniformity in detailing, easy form work, easy reinforcement rather than on the dimensions needed to carry vertical loads. Of course, buckling may restrict the height between floors or oblige the designer to increase column sections and wall thicknesses.

In high rises, design decisions are more critical. When buckling is excluded, maximum height of a column of any section, carrying its own weight, is given by:

$$h_{max} = \frac{\sigma_r}{n \rho g} \tag{6.3}$$

where ρ is the density of the material, σ_r the rupture stress and n imposed safety. For soft steel this height is 2750 meter, for normal concrete it is some 740 meter. As high rises approach that height, columns will occupy an ever larger part of the floor area at the lower storeys, leaving increasingly less net area to house floor functions. Very tall buildings thus become uneconomical structures, except if they are shaped with equal resistance, meaning that floor surfaces decrease more or less exponentially with height, see the John Hancock tower in Chicago and the Burj Khalifa tower in Dubai (Figure 6.5).

Figure 6.5. Shapes of equal resistance:
the John Hancock tower in Chicago and the Burj Khalifa in Dubai.

6.3.2 Horizontal load

6.3.2.1 Massive structures

Here, all load bearing walls have to carry the horizontal load. One way to get some indication of the forces exerted on each wall starts with positioning the floor plan in an $[x,y,z]$-axis system with x length, y width and z height. Floor slabs are assumed to be absolutely stiff horizontally. Suppose there are n load bearing wall entities (a wall entity consists of vertically stiff coupled walls). Each has as moment of inertia I_{xj} along the x- and I_{yj} along the y-axis ($1 \leq j \leq n$)). Each moment of inertia is represented by a vector in the centre of gravity of the entity considered. The vector field created that way has a centre of gravity $\bar{r}_o$, with ordinates x_o and y_o, called the 'stiffness centre'. A vector $\bar{r}_o$ links the origin of the $[x,y,z]$-axes to that centre (Figure 6.6).

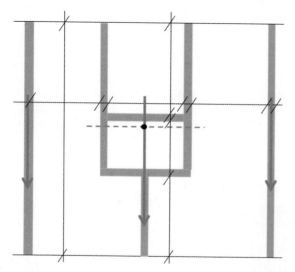

Figure 6.6. Massive structure: moments of inertia and stiffness centre of a 3 wall entities system.

Under horizontal load, each of the wall entities functions as a beam fixed in the foundations. When the resulting horizontal force passes through the stiffness centre, the stiff slab forces all entities to deflect identically. For a restrained beam with length L, deformation under a distributed load p is:

$$f = \beta \frac{p\,z^4}{E\,I} \tag{6.4}$$

with β a factor depending on load distribution and E the modulus of elasticity of the material used. Because f, β and z are identical for all entities, the following equations are valid for the case that all walls are constructed with the same material (E identical, $P = p \times h$ with h floor height in m):

$$P_{xj} = \gamma\,I_{xj} \qquad\qquad P_{yj} = \gamma\,I_{yj} \tag{6.5}$$

where γ is a factor of proportionality. The wind load component in x- and y-direction consequently splits up in as many horizontal forces as wall entities, or:

6.3 Structural system design

$$H_{wx} = \sum_{j=1}^{n}(P_{xj}) = \gamma \sum_{j=1}^{n}(I_{xj}) \qquad H_{wy} = \sum_{j=1}^{n}(P_{yj}) = \gamma \sum_{j=1}^{n}(I_{yj})$$

The factor of proportionality then becomes:

$$\gamma = H_{wx} \Big/ \sum_{j=1}^{n}(I_{xj}) = H_{wy} \Big/ \sum_{j=1}^{n}(I_{yj})$$

Or, each wall entity has to carry a horizontal load, equal to:

$$P_{xj} = H_{wx}\, I_{xj} \Big/ \sum_{j=1}^{n}(I_{xj}) \qquad P_{yj} = H_{wy}\, I_{yj} \Big/ \sum_{j=1}^{n}(I_{yj}) \qquad (6.6)$$

In case the resulting horizontal force does not pass through the centre of stiffness, a torsion moment is added to the force distribution just calculated:

$$M_w = H_{wx}\, y' + H_{wy}\, x' \qquad (6.7)$$

with x' and y' the distances along the x- and y-axis between the point of action of the resulting horizontal force (H_w) and the stiffness centre. For the components along the x- and y-axis Equation (6.7) simplifies to $M_{w1} = H_{wx}\, y'$ and $M_{w2} = H_{wy}\, x'$. The slabs now force the wall entities to deflect along the x- and y-axis, proportionally to the distance between their centre of gravity and the stiffness centre, or:

$$f_{xj} = \alpha\,(y_j - y_o) \qquad f_{yj} = \alpha\,(x_j - x_o)$$

The additional horizontal forces then equal:

$$P'_{xj} = \gamma\, \alpha\,(y_j - y_o)\, I_{xj} \qquad P'_{yj} = \gamma\, \alpha\,(x_j - x_o)\, I_{yj}$$

Moment equilibrium now imposes:

$$M_w = \sum_{j=1}^{n}\left[P'_{xj}(y_j - y_o) + P'_{yj}(x_j - x_o)\right]$$

which allows writing:

$$\gamma\,\alpha = M_w \Big/ \left\{ \sum_{j=1}^{n}\left[(y_j - y_o)^2\, I_{xj} + (x_j - x_o)^2\, I_{yj}\right] \right\}$$

Inclusion in the equations for the additional horizontal forces results in:

$$P'_{xj} = \frac{M_w\,(y_j - y_o)\, I_{xj}}{\sum_{j=1}^{n}\left[(y_j - y_o)^2\, I_{xj} + (x_j - x_o)^2\, I_{yj}\right]}$$

$$P'_{yj} = \frac{M_w\,(x_j - x_o)\, I_{yj}}{\sum_{j=1}^{n}\left[(y_j - y_o)^2\, I_{xj} + (x_j - x_o)^2\, I_{yj}\right]} \qquad (6.8)$$

Both equations underscore that the stiffest wall entities should be located farthest from the centre of stiffness. If not, the horizontal forces may overload the least stiff ones, subjecting the building to unacceptable torque.

This simple theory shows what is best and what should be avoided. Load bearing walls in one direction for example create problems. Stiffness in the other direction may be too small with unacceptable deflection as a result. In low and medium rises the solution is to stiffly couple walls and slabs so they act as frames (Figure 6.7).

In taller buildings correct stiffness presumes at least two stiff walls in one and one stiff wall in the other direction, positioned in a way their centre lines do not cross in a single point. The horizontal load component in the one direction is then carried by the two (Figure 6.8):

$$P_1 = H_{wy} \frac{I_{y1}}{I_{y1} + I_{y2}} \qquad P_2 = H_{wy} \frac{I_{y2}}{I_{y1} + I_{y2}}$$

while the one wall in the other direction carries the horizontal load component at that point whereas the two in that one direction neutralize torsion:

$$P'_{y1} = \frac{M_w (x_1 - x_o) I_{y1}}{\sum_{j=1}^{n} \left[(x_1 - x_o)^2 I_{y1} + (x_2 - x_o)^2 I_{y2} \right]} = P'_{y2} = \frac{M_w (x_2 - x_o) I_{y2}}{\sum_{j=1}^{n} \left[(x_1 - x_o)^2 I_{y1} + (x_2 - x_o)^2 I_{y2} \right]}$$

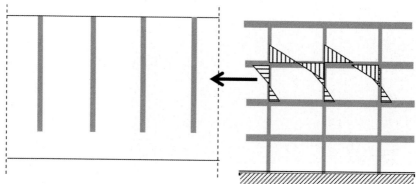

Figure 6.7. Walls and slabs acting as frames.

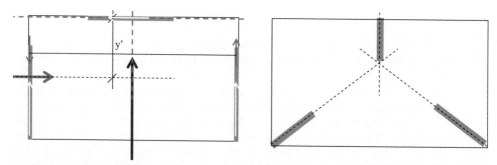

Figure 6.8. Three stiff walls, on the left correctly positioned, on the right wrongly positioned (centre lines crossing in one point).

6.3 Structural system design

When the centre lines of the three cross in one point, torsion cannot be neutralized, which of course is unacceptable (Figure 6.8).

6.3.2.2 Skeleton structures

In low rises, horizontal loads are easily withstood by framing beams and columns. Framing however does not prevent columns from experiencing quite significant deflections. These add per floor, causing a quick increase in displacement of the highest floor with the number of storeys. Once above 10 storeys, frames function properly only if the columns get wall-like dimensions in the load direction. That however diminishes freedom in floor organization, which is why for higher buildings preference is given to a combination of skeleton and stiff wall entities at locations that do not interfere with floor usage. This is the case for stairwells, elevator shafts, duct shaft, sanitary cores and facade strips. In high rises with height below 150 m it suffices to assemble them in one core, designed as beam fixed in the foundation (Figure 6.9).

Figure 6.9. High rise with central core under construction.

It is important that the core should be situated centrally in the building, or, when several cores are planned, they should be distributed as symmetrically as possible over the building's plan view. The stiffest cores or walls are best located farthest from the centre of stiffness. The columns, which must no longer withstand horizontal loads, can be designed hinged, carrying vertical loads only.

In tall buildings higher than 150 m, stiffness requirements gradually increase. For example if wind speed increases parabolically with height and displacement at the top is limited to a constant fraction of height, then the moment of inertia of the whole plan view has to increase with the fourth power of height. In extreme high rises under those conditions central cores will not suffice. The whole building should function as beam fixed in the foundations. That is realised by designing the envelope as a tube coupled to the central core at regular heights. Such structures are called tube in tube. Tubes are realized applying different structural solutions such as Vierendeel girders (Figure 6.10), mega trusses (Figure 6.11) or mega frames.

Figure 6.10. Tube as Vierendeel girder.

Figure 6.11. Trussed tube.

6.3.3 Dynamic horizontal loads

Dynamic horizontal loads not only engage structural stiffness, but also the building's mass and damping properties. The importance of stiffness and mass are clear looking to the resonance frequency (f_R) of a mass/spring system:

$$f_R = \frac{1}{2\pi}\sqrt{\frac{k}{m}} \qquad (6.9)$$

The stiffer a structure (k higher), the higher its resonance frequency. The more mass, the lower the resonance frequency. In theory, without damping, displacement at resonance is infinite. As damping increases, it drops quickly. But the acceleration, not the displacement is what hampers people, which is why ISO 6897 makes a link between resonance frequency and acceptable acceleration. At low frequencies, more stiffness may certainly conflict with acceleration limits set by the standard. In such cases more building mass or damping is a better option. Damping hardly has an effect on the resonance frequency but lowers acceleration considerably. Additional mass lowers both.

6.4 References and literature

[6.1]   Rosman, R. (1965). *Die Statische Berechnung von Hochhauswänden mit Öffnungsreihen.* Bauingenieur-Praxis, Wilhelm Ernst & Sohn, Berlin-München, 93 p. (in German).

[6.2]   Gerstle, K. (1967). *Basic Structural design.* Mc-Graw-Hill, 405 p.

[6.3]   Delrue, J. (1974). *Constructief Ontwerpen* (in Dutch).

[6.4]   ISO 6897 (1984). *Guidelines for the evaluation of the response of occupants of fixed structures, especially buildings and off-shore structures to low-frequency horizontal motion (0.063 to 1 Hz), ISO.*

[6.5]   Cziesielski, E. (1990). *Lehrbuch der Hochbaukonstruktionen.* B. G. Taubner Stuttgart, 627 p. (in German).

[6.6]   Fremantle, R. (1993). *Chicago.* Casa Editrice Bonechi, Firenze, 64 p.

[6.7]   *High-Rise Buildings.* Workshops at the TU-Delft, 1993–1997.

[6.8]   Van Oosterhout, G. (1997). Evaluation of the Voorhof II building refurbishment: a dynamic behaviour viewpoint. *Heron,* Vol. 42, no. 2, p. 97–111.

[6.9]   Werkstattbericht (1998). *Kohn Pedersen Fox.* DBZ 7/98, p. 49–56 (in German).

[6.10] The Museum of Modern Art, New York, 2003, Tall Buildings.

7 Floors

7.1 In general

Floors fulfil several functions. They parcel the building into usable volumes and surfaces. From that viewpoint, the net floor area is an important functional quality. They further transfer the vertical loads to the load bearing walls or skeleton columns and figure as stiff membranes, distributing horizontal loads over the different wall entities, stiff cores or columns. Floors also compartmentalize the building, as required for fire safety reasons.

Own weight, dead weight and live load deflect floors, quite an uneconomical way of load transfer as shown by the bending stress formula:

$$\sigma = \frac{M\,y}{I} \qquad (7.1)$$

with I the moment of inertia and y the distance to the neutral axis of the cross section. Only at a point farthest from that axis, is the material's strength fully engaged albeit only where the bending moment is maximal. Over the whole floor span, stress at the neutral axis remains zero. In view of efficient material use, best choice in terms of cross section is to locate as much material as possible far from that axis. But, a concentration at the top and bottom conflicts with shear strength, which requires sufficient material all over the cross section. Bending and shear combined then results in solutions such as I shaped steel sections, T shaped concrete beams and pre-stressed concrete hollow floor units. The fact that floors have to withstand bending also limits material choice. Enough tension strength is a prerequisite, no problem for timber, steel, reinforced concrete and pre-stressed concrete, but difficult for masonry where only vaulted solutions are possible. These turn bending into compression along the vault's centre line and lateral thrust at the vault springing.

Apart from the blind floor, additional layers compose the finished floor (Figure 7.1):

1. Ceiling. May be a plaster layer or a hung ceiling
2. Thermal insulation layer if needed
3. Screed. In case ducts, pipes and wiring must be housed in the floor's cross section, raised floor systems are a better choice
4. Floor covering (carpet, linoleum, synthetic floor cover, tiles, parquet, etc.)

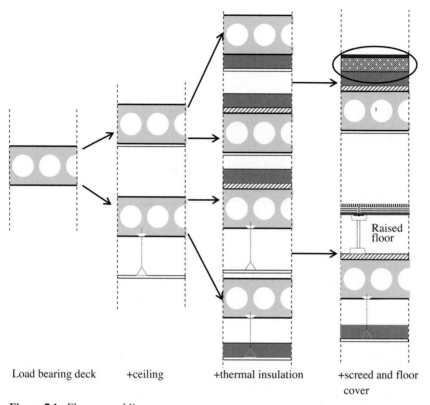

Figure 7.1. Floor assemblies.

7.2 Performance evaluation

7.2.1 Structural integrity

Structural integrity combines two requirements: safety against rupture and allowable deflection. Evaluation of the first may be done following two tracks: elastic design controlling stress and limit analysis, and limit analysis to investigate the load a floor deck is able to withstand without collapse. Elastic design assumes stresses can never pass linearity between stress and strain, keeping deformation reversible that way. Limit analysis instead considers safety against bending moments that form the number of plastic hinges needed for the floor deck to fail. Safety increases with the degree of static indeterminateness. Of course, deflections are no longer reversible.

Allowable deflection depends on function and floor finish. With hung ceilings or flexible finishes a deflection up to $L'/300$ is acceptable, L' being the effective floor span between supports. In case tiles are applied, deflection should be limited to $L'/500$. If parcelling the floor is done using movable partitions then $L'/750$ is advised. A larger deflection complicates partition mounting and may create sound paths at floor and ceiling.

7.2 Performance evaluation

Elastic deflection is given by:

$$f = \beta \frac{p L^4}{E I} \tag{7.2}$$

With $f \leq L/n$ and $300 \leq n \leq n_{\text{allowed}}$, the relation between the moment of inertia I and span L becomes $I \div L^3$. Allowable stress design gives $I \div L^2$. As a consequence, for larger spans, not stress but deflection defines the cross section. In case the floor has to carry vibration sensitive equipment, dynamic response must be checked (resonance frequency, vibration amplitudes). The same holds for varying live loads, as on a dance floor. The ultimate deflection anyhow depends on creep.

7.2.2 Building physics: heat-air-moisture

7.2.2.1 Air tightness

Deficient air-tightness limits airborne and contact sound insulation and degrades the floor's fire resistance. Massive floors do not pose much of a problem, except at penetrations for electricity, piping and ducts, where adequate attention must be given to correct caulking, for example by filling with mineral wool and finishing with a seal.

In case hollow structural floor units are used, not closing the hollows at the envelope support allows outside air to traverse the slab, causing low floor temperatures and increased heat loss. Prefabricated reinforced or pre-stressed concrete hollow structural floor units, supported by a cavity wall, are an example. The open hollows may collect cavity side rainwater runoff, causing dripping indoors where floor deflection is maximal (Figure 7.2).

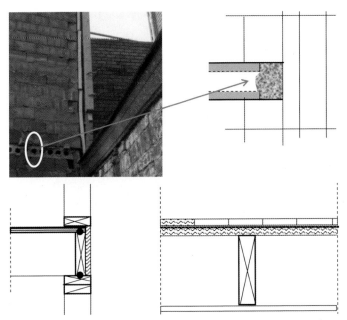

Figure 7.2. Closing the hollows in concrete structural floor units at the envelope support; timber floor with double deck, plastic foil in between, airtight ceiling.

With timber floors, not only the supports must be air tightened but also the load bearing deck should be detailed in a way air tightness is guaranteed, for example by using double-laid plywood or OSB-sheaths with in between a plastic foil or, by installation of an airtight hung ceiling afterwards (Figure 7.2).

7.2.2.2 Thermal transmittance

For storey floors and floors above outdoors, whole and clear floor thermal transmittance are given by:

$$U = U_o + \frac{\sum(\psi_j L_j) + \sum \xi_k}{A} \quad \text{with} \quad U_o = \frac{1}{1/h_1 + \sum_{j=1}^{n} R_j + 1/h_2} \quad (7.3)$$

and:

Case	h_1 W/(m²·K)	h_2 W/(m²·K)
Floor above outdoors	6	25
Storey floor, heat flowing downwards	6	6
Storey floor, heat flowing upwards	10	10

In these formulas, U is the whole floor thermal transmittance with ψ linear thermal transmittance of the linear thermal bridges, if any, L their length and ξ the local thermal transmittance of local thermal bridges, if any. U_o stands for the clear floor thermal transmittance. Thermal bridging at facade/floor supports however is typically attributed to the envelope.

Example

Assume following floor assembly (from bottom to top):

Layer	Thickness cm	Thermal conductivity W/(m·K)	Thermal resistance m²·K/W
Gypsum plaster	1.5	0.28	0.054
Prefab floor elements	14		0.150
Screed	6	1	0.060
Tiles	1	1.5	0.007
		Sum	0.27

Clear floor thermal transmittance:

Case	U W/(m²·K)
Floor above outdoors	2.09
Storey floor, heat flowing downwards	1.65
Storey floor, heat flowing upwards	2.12

7.2 Performance evaluation

The result shows that (1) the clear floor thermal transmittance of non-insulated floors scores high, (2) its value differs considerably depending on the case and heat flow direction.

A first check serving as a basis for requirements is foot comfort. In residential spaces, floor temperature should not drop below 19 °C during very cold days. This is not a problem for storey floors between heated spaces but for floors above outdoors and storey floors above non-heated, strongly ventilated spaces, these 19 °C require the clear floor thermal transmittance to stay below:

$$U_o \leq 6 \frac{\theta_i - 19}{\theta_i - \theta_e} \tag{7.4}$$

In case the operative temperature indoors is 21 °C and outdoors one has the design temperature, the formula gives the values in Table 7.1:

Table 7.1. Floor above outdoors, clear floor thermal transmittance threshold for foot comfort.

θ_d °C	U W/(m²·K)
0	0.57
−8	0.40
−10	0.38
−12	0.36
−18	0.31
−25	0.26
−30	0.24

A second check concerns energy efficiency. Floors above outdoors and storey floors above non-heated, well-ventilated spaces should have whole floor thermal transmittances close to the life cycle cost optimum. That optimum of course differs between countries, depending on climate, energy prices and construction costs:

Country	Case	$U \leq$ W/(m²·K)
Belgium	Floor above outdoors	0.30 (from 2012 on)
	Floor between apartments	1.00
Canada (Quebec)	Floor above outdoors	0.28 (1999)
Denmark	Floor above outdoors	0.12–0.30 (2007)
Germany	Floor above outdoors	0.40 (2007)
Finland	Floor above outdoors	0.25 (2007)
France	Floor above outdoors	0.27 (2007)
The Netherlands	Floor above outdoors	0.37 (2007)
Austria	Floor above outdoors	0.35 (2007)
UK	Floor above outdoors	0.25 (2007)

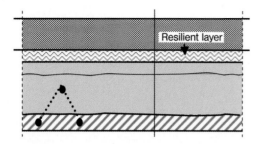

Figure 7.3. Screed with 3 cm thick resilient layer below.

Table 7.2. Floor above outdoors, insulation layer.

Insulation material	Thickness m		
	$U = 0.4$ W/(m²·K)	$U = 0.2$ W/(m²·K)	$U = 0.10$ W/(m²·K)
Mineral wool	0.09	0.18	0.36
EPS	0.09	0.18	0.36
XPS	0.08	0.16	0.32
PUR	0.06	0.13	0.26

For low energy buildings, 0.2 to 0.25 W/(m²·K) is the value considered, whereas passive buildings advance $U < 0.15$ W/(m²·K). The following floor assemblies meet the requirement of a clear floor thermal transmittance ≤ 1 W/(m²·K) and ≤ 0.4 W/(m²·K):

- $U \leq 1$ W/(m²·K)
  - Floating screed with 3 cm thick resilient layer below (Figure 7.3).
  - Light-weight screed with thermal resistance above 0.65 m²·K/W (for example 5 cm polystyrene concrete with density 250 kg/m³ or less)
- $U \leq 0.4$ W/(m²·K)
  Insulation layer below the blind floor or between blind floor and screed. Table 7.2 gives the thicknesses needed for the example above. The table underlines insulation thickness increases steeply with lower thermal transmittances, though added energy efficiency becomes gradually more marginal.

7.2.2.3 Transient response

Dampening of sol-air temperature harmonics is a performance requirement at room level. Yet, building parts with high admittance and high dynamic thermal resistance ameliorate room response, with floors having more impact than partition walls. Indeed, the solar spot moves over the floor most of the day. Part of that radiation is absorbed and released with a time lag. As saturated colours augment short wave absorption, heat storage increases the darker the floor's finish, the higher its contact coefficient, the heavier its weight and the farther away the insulation layer from the radiated surface. As the admittances listed in Table 7.3 underline, assembly (1) of Figure 7.4 performs better than (2), which in turn is better than (3).

7.2 Performance evaluation

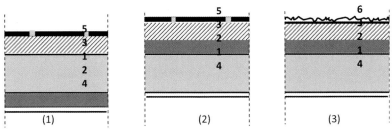

1. Reinforced concrete slab
2. Thermal insulation (EPS)
3. Screed
4. Plastered ceiling
5. Tiled finish
6. Carpet

Figure 7.4. The floor assemblies (1), (2) and (3).

Table 7.3. Admittances of the floors of Figure 7.4 (surface resistance $1/h_i$ not considered).

Floor assembly	Admittance			
	1 day period		1 week period	
	Ad W/(m²·K)	Time shift hours	Ad W/(m²·K)	Time shift hours
(1)	9.9	2.08	3.7	30.2
(2)	7.5	2.20	1.2	31.5
(3)	4.1	2.62	0.95	26.4

7.2.2.4 Moisture tolerance

Floors struggle with building moisture, rising damp, rain, hygroscopic moisture, surface condensation and interstitial moisture deposit.

Building moisture

In cast concrete slabs most of the mixing water becomes building moisture. Of course, independent of slab type, a fresh cement-based screed always shows above critical wetness. Applying a vapour retarding, joint-less floor finish like linoleum or any other synthetic material should therefore wait until the screed reaches hygroscopic equilibrium with indoors. As this takes too much time according to many practitioners and principals, too early finishing is current practice albeit this may have nasty consequences. Consider the situation of Figure 7.5. Once the screed is cast, heating and ventilating the storeys above and below intensifies upward vapour diffusion, which in turn shortens the time (t) before the second drying phase starts:

$$t = \frac{d}{g_{vd}}\left[w_o - \left(w_{cr} + \frac{g_{vd}\, d}{3\, D_w}\right)\right] \quad \text{met} \quad g_{vd} = \beta\, p_{sat}\,(1-\varphi)$$

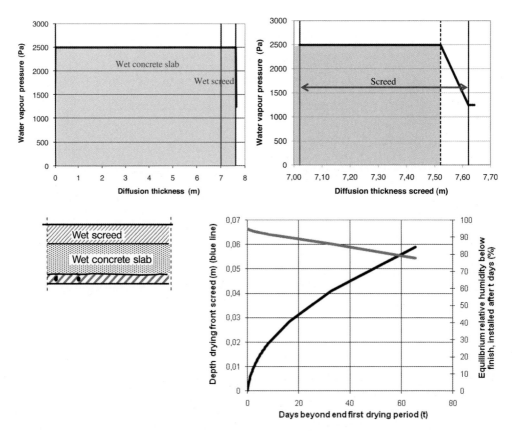

Figure 7.5. Moist screed. Air temperature above and below 21 °C, relative humidity 50%. Vapour tight finish installed too early. Upper graph left gives water vapour pressure in the fresh screed, upper graph right in the screed with 1 cm thick air-dry upper layer. Lowest graph shows how relative humidity below the finish (top line) reaches a new equilibrium for a given thickness of that dry layer (bottom line). (μd_{screed} = 0.6 m, μd_{finish} = 50 m).

In that formula w_o is the building moisture content in the screed, w_{cr} its critical moisture content, d thickness, D_w moisture diffusivity, g_{vd} drying rate, p_{sat} vapour saturation pressure and ϕ relative humidity in the air above. Clearly, doubling the drying rate halves that time. The price paid however is higher average moisture content in the screed at that moment. Yet, as soon as the screed surface looks dry, the floor layer finishes it. At that moment, surface and air relative humidity are the same but in the screed, 100% is still noted at the drying front. Once the finish is glued, the moisture in the screed redistributes, moving the relative humidity below the finish back to high values, see Figure 7.5.

Incident solar radiation now can lift vapour pressure below high enough to initiate blistering, followed at night by screed moisture condensation in the blisters. That process ends with dissolving of moisture sensitive finish glue, growing blisters and degraded floor usability. Avoidance demands postponing floor finishing until the whole screed has reached hygroscopic equilibrium, choosing floor finishes with low vapour resistance (how low is case dependent) and/or using moisture resistant glue.

7.2 Performance evaluation

Rising damp

Rising damp can moisten floors when bottom-wet partitions and outer walls contact floor materials with higher suction potential (Figure 7.6). Cast concrete is such a material, albeit moisture uptake may be too slow to cause problems. Masonry vaults and timber beams instead can suck quite some moisture. The same holds for screeds. Besides, horizontal capillary movement never stops. It takes time, but screeds sometimes get capillary wet all over their surface.

Figure 7.6. Floor in contact with bottom-wet partition wall.

Wind driven rain

Wind driven rain risk is low, though there are exceptions. In timber floors, purlin heads may turn wet together with the supporting facade masonry, causing fungal attack and rot. As already mentioned, hollow structural floor units contacting the air space in a cavity wall without or with erroneously mounted trays, may see cavity side run-off collect in the hollows and drain to the most deflected floor zones where the water can drip in the room below. Prevention demands closing the hollows at the floor support with cast concrete and inserting correct trays above (Figure 7.7).

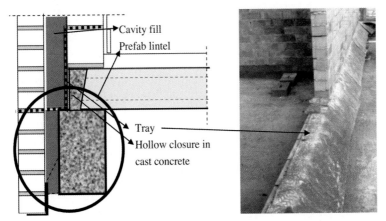

Figure 7.7. Floor supported by a lintel in a cavity wall: closing the floor hollows with cast concrete, inserting a correctly mounted tray.

Hygroscopic moisture

When the monthly mean relative humidity at the coldest floor spot exceeds 80%, mould probability approaches 1 there. Keeping probability below 0.05 demands temperature factors (f_{hi}) above 0.7 in moderate climates. Especially at thermal bridges but also with structural floor units that have unclosed hollows at the envelope supports, values above 0.7 may be hard to realize.

Avoiding dust mite overpopulation in floor carpets imposes even more severe temperature and relative humidity conditions. For dermatophagoides pteronyssimus (a wide spread dust mite species) critical long-term upper treshold relative humidity is (above, dust mite population starts to explode):

Temperature (°C)	15	25	35	45
Relative humidity (%)	52	58	63	69

Surface condensation

With temperature factors above 0.7, surface condensation probability at design temperature also remains below 0.05.

Interstitial condensation

Interstitial condensation risk exists when temperature and vapour pressure at both sides differ.

Storey floors

For storey floors, temperatures above and below are usually close, making interstitial condensation highly unlikely. Of course, tolerance for building moisture, rain, hygroscopic moisture and surface condensation must be guaranteed when designing and constructing the floor.

Floors above outdoors

The floor now separates indoors, where set-point temperatures guarantee comfort and building usability, from the weather outdoors, be it without sun, under-cooling and rain. In many climates, temperature differences between both are large enough to turn interstitial condensation into a risk. How large that risk is and how severe the consequences are, depends on the presence or absence of building moisture, the indoor climate class (ICC 1, 2, 3, 4 or 5), location of the thermal insulation in the floor assembly and diffusion resistance of the layer at the cold side, the outside in cold and cool climates, and the inside in hot and humid climates. Happily, massive floors rarely experience air in- or exfiltration. No building moisture is realistic for timber floors without screed. Thermal insulation can be located between the blind floor and the screed or below the blind floor. With massive slabs, the insulation in-between demands a levelling layer. With timber floors, the insulation in-between means placing it on a plywood or OSB subfloor. The insulation below may get different finishes: plaster on metal wire, timber lathing or fibre-cement boards. The basic questions are: does the floor need a vapour retarding layer, and if yes, where to put it and what quality class required?

7.2 Performance evaluation

(1) Thermal insulation between blind floor and screed

To prevent cement from penetrating the joints between the insulation boards, these are covered with a 0.2 mm thick polyethylene foil before casting the screed. Such foil is highly vapour retarding, reducing to zero problematic interstitial condensation risk in the insulation, the levelling layer and the plywood or OSB subfloor, first from building moisture from the screed and later of water vapour released indoors.

(2) Thermal insulation below the blind floor

Analysis for a cool climate shows that even in humid buildings (climate class 5 according to EN ISO 13788) interstitial condensation of building moisture from the screed or water vapour released indoors is prevented on condition the following ratio between vapour diffusion thickness of the outside finish and the floor assembly above is observed:

$$[\mu d]_{\text{finish}} \leq \left(\sum [\mu d]_{\text{floor above}} \right) / 15 \tag{7.5}$$

That ratio does not apply in hot and humid climates. There, the diffusion thickness of outside finish and insulation must pass that of the blind floor and the layers on top. But, let us return to cool climates. Take the floor assembly of Figure 7.8.

Assume a monthly mean inside temperature from 17 °C in winter to 23 °C in summer and as vapour pressure indoors the threshold between ICC 3 and ICC 4. The concrete floor slab is air-dry, the day the screed, containing 12 litres of building moisture per m², is cast. Material properties:

Layer	D cm	λ W/(m · K)	μ –
Floor cover	1	1.5	200
Screed	6	1	10
Concrete slab	14	2.6	50
Thermal insulation	7.5	0.035	1.2
Outside finish	1.5	0.3	Variable

If the outside finish has the diffusion thickness ratio of [7.5], Glaser's method for January gives the steady state vapour saturation and vapour pressures of Figure 7.8a (outside temperature 2.7 °C, outside vapour pressure 675 Pa). Although relative humidity at the backside of the finish nears 100%, no building moisture condenses there. Screed drying in turn progresses as shown in Figure 7.8b. Figure 7.8c gives the result in case one does not respect that ratio. As long as the screed contains building moisture, condensate accumulates at the backside of the outside finish (see Figure 7.8d).

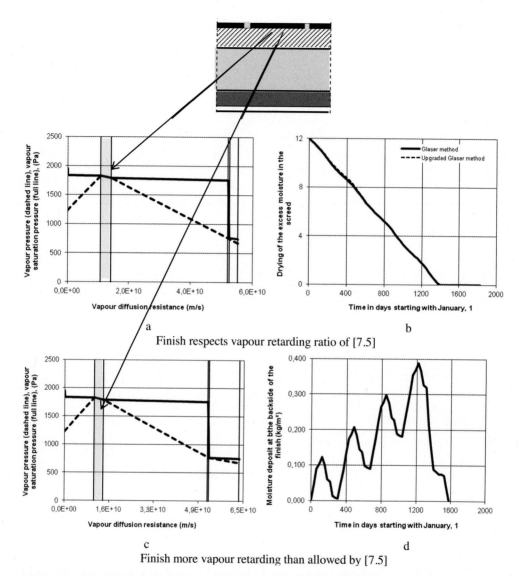

Figure 7.8. Floor, building moisture in the screed, no vapour retarder (MDRY year for Uccle, Belgium).

7.2 Performance evaluation

7.2.2.5 Thermal bridging

Most critical is the floor support at the outer walls. If incorrectly designed, thermal bridging is manifest, see Figure 7.9.

Detail (a) shows this for a concrete slab touching the veneer of a cavity wall filled with 8 cm mineral fibre. A linear thermal transmittance 0.55 W/(m · K) is far from negligible, while the temperature ratio at the ceiling edge drops below 0.7, the pivot from negligible to a mould risk of almost 1. Things turn even worse when the insulation stops above the cavity tray, a common flaw, see detail (c). A correct design is nevertheless simple. The slab must end at the interface between inside leaf and fill. That way the fill acts as thermal cut, see detail (b). Of course, when casting the slab, one needs edge formwork, whereas in detail 7.9a the veneer acts as such, although the cavity must still be covered before casting the slab. Figure 7.6 above showed how to detail the floor slab support at lintels.

Because, as already mentioned, thermal bridging mainly concerns the facade support, related linear and local thermal transmittances are typically attributed to the envelope. In that sense, whole thermal transmittance of most floors (U) equals the clear value (U_o). However, beam floors with insulated hollow tiles may see their whole floor thermal resistance reduced substantially compared to what the insulation could offer – see Table 7.4.

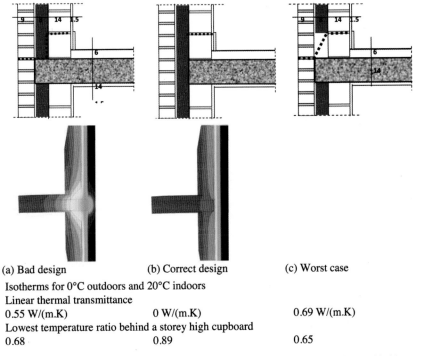

(a) Bad design　　　　　　(b) Correct design　　　　(c) Worst case
Isotherms for 0°C outdoors and 20°C indoors
Linear thermal transmittance
0.55 W/(m.K)　　　　　　　0 W/(m.K)　　　　　　　0.69 W/(m.K)
Lowest temperature ratio behind a storey high cupboard
0.68　　　　　　　　　　　0.89　　　　　　　　　　0.65

Figure 7.9. Floor support along the facade as an example of a potential thermal bridge: (a) as it should not be designed, (b) correct design, (c) worst case with the insulation first starting above the cavity tray.

Table 7.4. Projected clear and effective whole floor thermal resistance of beam floors with hollow tiles.

Assembly	Projected clear floor thermal resistance $m^2 \cdot K/W$	Effective whole floor thermal resistance $m^2 \cdot K/W$
Hollow tiles not full-foamed with PUR	0.25	0.15
Hollow tiles full-foamed with PUR	1.65	0.20
Hollow tiles not full-foamed with PUR	0.30	0.20
Hollow tiles full-foamed with PUR	1.85	0.30
Hollow EPS/wood wool cement tiles, 15 cm high	2.0	0.55
Hollow EPS/wood wool cement tiles, 20 cm high	3.1	0.60
Hollow EPS tiles, 16 cm high, with EPS lips covering the concrete joists		1.50

7.2.3 Building physics: acoustics

7.2.3.1 Airborne noise

For storey floors separating residential units, the EN standard demands a transmission loss $R_{w,500} \geq 52$ dB(A). That is a minimum. To really exclude complaints, $R_{w,500} \geq 60$ dB(A) is needed. With massive floors, weights above 400 kg/m² suffice to reach 52 dB(A), demanding a 14 cm thick concrete slab, 6 cm thick cement screed and 1.5 cm gypsum plaster as ceiling finish. Attaining 60 dB(A) requires more than double that weight or a smart use of composite assemblies with a floating screed. With lightweight floors, a double assembly consisting of an airtight ceiling with own load-bearing structure and sound absorbing layer above, mounted below the timber floor offers a way out on condition the timber floor is air-tight.

7.2.3.2 Impact noise

To judge impact noise insulation, the ISO standard compares insulation of the assembly forwarded with a reference curve allowing a one number evaluation: the difference at 500 Hz between the reference curve and the impact noise insulation demanded. The performance requested typically lies between 0 and 5 dB at 500 Hz. But, every measure which upgrades airborne transmission loss, also ameliorates impact noise insulation, though the inverse is not true. The principle offers three possibilities to upgrade impact noise insulation (see Table 7.5, the values given are measured impact noise at 500 Hz, not the difference with the ISO-curve at 500 Hz):

7.2 Performance evaluation

- Floating screed. Works well when the load-bearing slab is heavy. Above 350 kg/m², dynamic resilience of the elastic layer is less critical. Below 350 kg/m², more severe dynamic resilience values are demanded:

Slab weight slab kg/m²	Dynamic resilience N/m³
≤ 350	< 30
> 350	$30 \leq s \leq 90$

- Thick, heavy load-bearing slab (for example, a concrete slab with thickness ≥ 20 cm), finished with a floor cover having a high improvement value for contact noise
- Double floor assembly

Table 7.5. Impact noise for different floor assemblies, values at 500 Hz.

Assembly	Impact noise $L_{n,500}$ dB
1. Timber floor, top down: • 22 mm chipboard • Joists 240 × 80 mm, centre to centre 60 cm, 100 mm mineral wool in between • 2 × 12.5 mm gypsum board screwed on an elastically hung timber frame	73
2. Timber floor, top down: • 22 mm thick chipboard • Joists 220 × 55 mm, centre to centre 60 cm, 100 mm mineral wool in between • 12.5 mm gypsum board screwed on a timber frame composed of laths 48 × 24 mm	77
3. Concrete slab, 140 mm thick	76
4. As 1, however with a 19 mm thick chipboard floating floor on 30 mm thick mineral wool boards	64
5. As 2, however with a 50 mm thick cement floating floor on 30 mm thick resilient EPS, plus extra layer of 9.5 mm gypsum board against the 12.5 mm already mounted	57
6. As 3., however with a 50 mm thick cement floating floor on 30 mm thick elastic EPS	50

7.2.4 Durability

Storey floors hardly experience temperature variations. Reinforced concrete slabs therefore only suffer from chemical and drying shrinkage. In large floor surfaces, neutralizing chemical shrinkage requires adapted casting schemes.

In floors above outdoors, the insulation layer's location defines how large the temperature variations in the load-bearing slab will be. If located between screed and blind floor, the last has to buffer the whole variation in outside temperature. Mounting it below the blind floor instead largely excludes the outside temperature as a factor of influence, see Table 7.6.

Table 7.6. Annual temperature variation in the blind concrete floor of assemblies (1) and (2), see Figure 7.4 (Uccle climate, indoors 17 °C in winter, 23 °C in summer).

Floor assembly (see Figure 7.4)	Annual temperature variation °C		
	Lowest value	Highest value	Difference
(1)	13.8	23.1	9.3
(2)	−14	27.1	41.1

7.2.5 Fire safety

Storey floors separate fire compartments. For medium and high-rise buildings, performance requirement for compartment separations is a fire resistance passing 90′ for all three aspects to be considered: safety against collapse, smoke tightness, and surface temperature at the other side. The floor assembly must also consist of non-combustible materials. In low-rises, a fire resistance of 30′ is allowed, without the requirement to use only 'non-combustible materials'. The warming curve of a floor under one-sided fire may be approximated as:

$$\theta_{fl} = \theta_\infty + (\theta_o - \theta_\infty) \exp\left(-\frac{t}{\tau}\right) \qquad \theta_{reinf} = \theta_{fl} + (1193 - \theta_{fl})\frac{R^o_{reinf}}{R_1}$$

with:

$$\tau = \frac{R_1 \, R_{2,eq} \, \rho \, c \, d}{R_1 + R_{2,eq}} \qquad \theta_\infty = \frac{\theta_f \, R_{2,eq} + \theta_e \, R_1}{R_1 + R_{2,eq}}$$

$$R_{2,eq} = R_2 + \frac{A_{fl}}{A_{fac} \, U_{m,fac} + 0.34 \, n \, V} \qquad R^o_{reinf} = d^o_{reinf} / \lambda_{concrete}$$

where θ_o is the temperature of the floor before the fire started, θ_f the temperature at the fire side (the underside, equals 1193 °C), θ_e the outside temperature (10°), θ_{fl} the temperature in the thermal centre of gravity in the floor, R_1 the thermal resistance between the fire side and that thermal centre of gravity in the floor, $R_{2,eq}$ the equivalent thermal resistance between that thermal centre of gravity and the ventilated zone at the other side, R^o_{reinf} the thermal resistance and d^o_{reinf} the distance between the thermal centre of gravity and the reinforcement in the concrete slab, $\rho \, c \, d$ the thermal capacity of the floor and A_{fl} floor area. A_{fac} is the area and $U_{m,fac}$ the mean thermal transmittance of the facade enclosing the zone at the other side, V that zone's volume and n ventilation rate there (all in SI-units).

Figure 7.10 shows which measure increase fire resistance the most. To simplify calculations, the fire curve was reduced to a step function from 21 °C to 1193 °C (θ_f), whereas the reinforcement was considered perfectly plastic beyond 580 °C. All other system properties, including the surface film resistance at the fireside (±0.0046 m² · K/W) were kept constant. Without insulation, the bars reach 580 °C after 200′. Increased heat capacity of a thicker concrete slab hardly changes things. Mounting a fireproof, airtight hung ceiling with 75 mm mineral wool above is much more effective. As long it does not collapse, the bars warm up quite slowly.

7.3 Design and execution

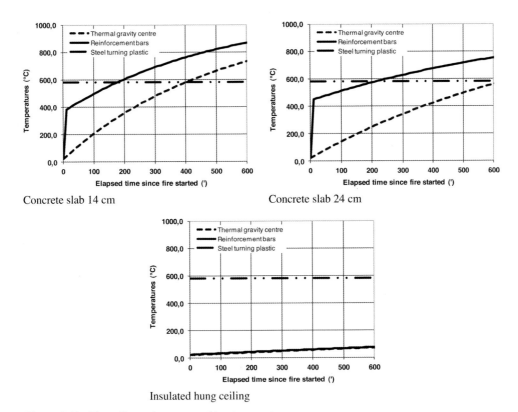

Figure 7.10. Floor, fire resistance: steel bar temperature.

7.3 Design and execution

7.3.1 In general

Floors, included those above basements and crawlspace and in basements with multiple levels, are classified according to the material used and the span, see Table 7.7.

Table 7.7. Floor construction, an overview.

Material	Span	
	≤ 6 m	> 6 m
Timber (not in basements with multiple floors)	Joists	Laminated beams
Reinforced and pre-stressed concrete	Slabs Prefabricated structural floor units	Ribbed slabs Cassette slabs Slab and beam Beamless slabs
Steel	Steel/concrete composite slabs	Beams and girders

7.3.2 Timber floors

7.3.2.1 Span below 6 m

For spans below 6 m, timber floors are constructed using joists with as boarding 10 to 22 mm thick plywood, OSB or, in older buildings, tongue and groove timber planks. The joist cross sections as the distances centre to centre are standardized:

European coniferous wood	Sections	63/170, 75/195, 75/220 mm
	Distances centre to centre.	30, 40, 60 cm
Canadian coniferous wood	Sections	60/180, 80/200, 80/240, 80/300 mm
	Distances centre to centre.	30, 40, 60 cm

As a rule, deflection determines sections. In technical specifications one may, given the span and the load per m², find graphs giving the section and distance centre to centre to limit deflection to 1/300 of the span and tensile stress to for example 8.5 MPa. Figure 7.10 gives an example, G being dead weight and Q live load, both in N/m². Usage of the graphs is obvious. The documents also give boarding thickness, see Table 7.8 for plywood.

To avoid lateral buckling joist are strutted at mid-span (Figure 7.12a). Openings for staircases and chimneys are formed by header joists supported by doubled joists. The correct header sections are important. If needed, headers are doubled (Figure 7.12b). Joists must be coupled

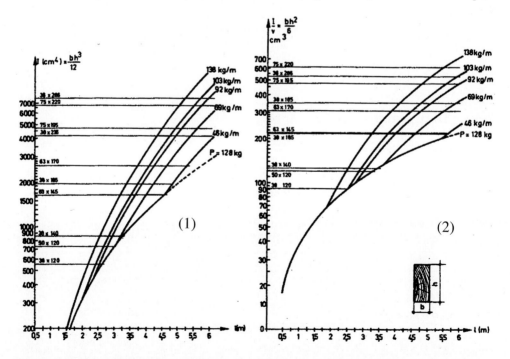

Figure 7.11. Timber floors, joists. Graph (1) gives the moment of inertia needed as function of span and load per meter run; graph (2) does the same for the section modulus.

7.3 Design and execution

above all load-bearing partitions (Figure 7.12c). Non-bearing masonry partitions perpendicular to the span act as local loads whereas non-bearing ones parallel to the joists have to be supported by a reinforced concrete beam (Figure 7.12d). If for any reason this is not possible, lightweight partitions should be used.

Table 7.8. Boarding thickness for plywood.

Thickness (mm)	10	13	16	19	22
Load	Span between joists in cm				
$G + Q = 2500$ N/m^2	42	55	67	80	97
Local load of 2000 N	28	43	62	74	97

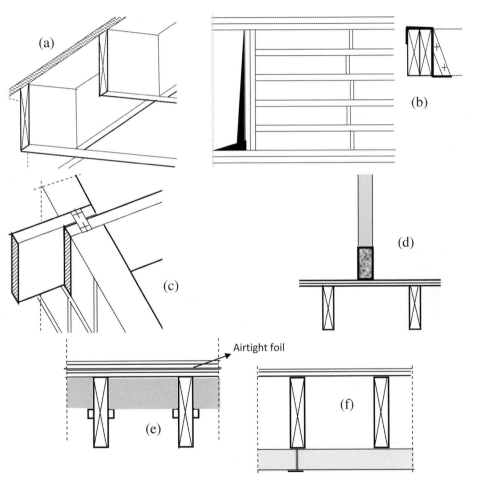

Figure 7.12. Timber floors, (a) strutting joists at mid-span, (b) header joists supported by doubled joists, (c) joists coupled above load-bearing masonry walls, (d) non-bearing masonry walls parallel to the span, (e) insulation with mineral wool, (f) insulation with rigid foam boards.

Delicate from a building physics point of view are:

- Air-tightness
  How to achieve has been explained.
- Transient response
  Volumetric heat capacity is negligible.
- Moisture tolerance
  Avoid joist supports in masonry outer walls which become wet.
- Sound insulation
  If an airborne noise transmission loss above 40 dB(A) and acceptable impact noise insulation are the objective, an airtight floor/airtight ceiling combination with sound absorbing boards in between gives relief. Also adding weight may help.
- Fire safety
  Finish timber floors with a non-combustible, well insulated hung ceiling and a non-combustible floor cover.

Adding thermal insulation is best done as indicated in Figure 7.12e and f. Mineral wool boards should be pressed against the subfloor while rigid foam boards are best mounted against the joist's underside and air-tightened by caulking or taping the joints in between. Air tightening a floor insulated with mineral wool can be done by mounting an additional layer of rigid foam boards the way just described.

7.3.2.2 Spans above 6 m

In such cases, laminated timber beams are used as main girders and timber joists as secondary beams.

7.3.3 Concrete slabs and prefabricated structural floor units

7.3.3.1 Span below 6 m

- Reinforced concrete slabs
  As a rule of thumb, slab thickness should equal $L'/35 + 3.5$ (cm), with L' effective span in cm and 3.5 cm the concrete cover from the middle of the steel bars. For the definition of the effective span, see Figure 7.13.

7.3 Design and execution

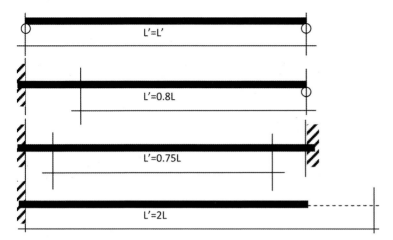

Figure 7.13. Reinforced concrete slabs: effective span L'.

- Precast wide slabs with on site casted concrete top layer
  The precast wide slabs that contain the steel bars needed to bridge the span act as formwork. Before casting the top layer, steel bars orthogonal to the span and steel bars above load-bearing walls are added (Figure 7.14a).
- Prefabricated structural floor units
  Typically, they consist of hollow concrete floor units, containing the necessary steel bars or post-stressed. The activity on the building site consists of laying the units side by side and filling the joints in between with micro-concrete. Lay out plans are typically drawn and specific problems solved by the manufacturer (Figure 7.14b).
- Beam floors with hollow tiles
  In such floors slabs are replaced by reinforced or post-stressed concrete beams with hollow tiles in between and cast concrete on top (Figure 7.14c). The hollow tiles are made of lightweight concrete, clay or expanded polystyrene. The last acts as floor insulation, if designed with lips covering the beams. To facilitate design, manufacturers publish tables with effective spans as function of dead weight and live load for each type produced.

As concrete slabs, precast wide slab floors, prefabricated structural floor units and beam floors with hollow tiles have to meet all building physics performances, thickness should be chosen with care. Often fire resistance and sound insulation are more demanding than strength and stiffness. Further on, blind floors must be finished in a way they guarantee the thermal transmittance required, achieve required impact noise insulation, can be walked-on without slipping, etc. The additional layers include levelling, if necessary, thermal insulation or a resilient layer, the screed, a plastered or an airtight fire proof hung ceiling and the floor cover.

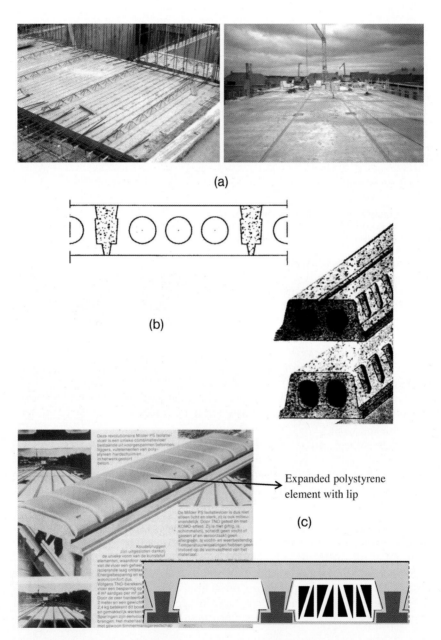

Figure 7.14. From top to bottom: (a) precast wide slabs with on-site casted concrete top layer, (b) prefabricated structural floor units, (c) beam floors with hollow tiles.

7.3 Design and execution

7.3.3.2 Span above 6 m

For spans above 6 meters, the solutions mentioned become uneconomical, mainly because of the too large thicknesses needed. Better choices then become:

- Beamed floor
  The structure is parcelled by a system of main and secondary beams with one of the solutions discussed under spans less than 6 meter for the floor. If precast wide slabs form the floor, then the main and secondary beams act as T girders (Figure 7.15a).
- Ribbed floor
  Here the distance centre to centre between the secondary beams is so small the floor slab functions as a system of T beams. The slender flanges, called ribs, have a height equal to $\approx 1/20^{th}$ of the effective span, whereas the concrete deck is typically 10 cm thick or more (Figure 7.15b). The disadvantage of a ribbed floor is the complex formwork, which is why prefabricated formworks were developed, ribs are prefabricated and the floor is constructed, using precast wide slabs. An alternative are prefab T, TT or U-units.

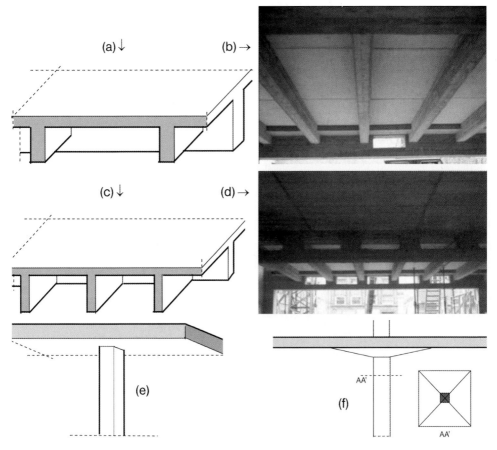

Figure 7.15. Concrete floors with span beyond 6 m:
(a) beamed, (b) ribbed, (c) cassette, (d) Vierendeel girders, (e) beamless, (f) mushroom floor.

- Cassette floor
  In case the column raster is close to foursquare, ribs in both orthogonal directions make sense. The result is a cassette floor. Also here prefabrication is welcomed (Figure 7.15c).
- Beamless floor
  Disadvantage of beamed, ribbed and cassette floors is the considerable construction height, minimizing the space left for piping and ductwork (HVAC, sanitary, gas, sprinklers, electricity). The main beams also obstruct easy duct layout, except if designed as Vierendeel girders (Figure 7.15d). Beamless floor systems solve that problem, on condition the column raster is close to foursquare. They transmit loads in two directions while experiencing important punching forces around columns. These define slab thickness and shear reinforcement (Figure 7.15e). If necessary, mushroom floors are used (Figure 7.15f).

7.3.4 Steel floors

7.3.4.1 Span below 6 m

Steel is not suited to construct floors. Nonetheless, during the past decade, composite steel/concrete solutions have gained market share with ribbed steel panels as formwork and reinforcement for the on-site casted concrete deck (Figure 7.16a). A drawback is building moisture cannot dry via the slab's underside. A vapour tight finish will then keep concrete and screed wet with inconveniences like blisters in the finish, corrosion, dripping moisture once the steel panels show corrosion pits.

7.3.4.2 Span above 6 m

Steel excels as material for beams. A simple floor solution is I-girders as main and secondary beams, with a concrete or steel/concrete deck above. Such a system can be optimized in terms of material usage and height by using shear connectors to activate steel/concrete interaction.

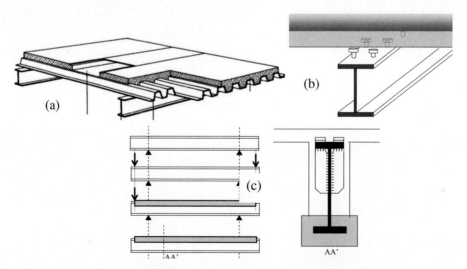

Figure 7.16. Steel floors: (a) steel/concrete composite deck, (2) steel girders/concrete deck with shear connectors, (c) preflex girder.

That results in composite sections with the steel under tension and the concrete under compression (Figure 7.16b). Preflex beams are realized by preloading a steel I-girder and casting concrete around the flange under tension. After the concrete is set the load is removed, resulting in beams with pre-stressed concrete around that flange, largely increasing load capacity and stiffness (Figure 7.16c).

As with concrete floor decks, building physics related performances demand additional layers. For example for fire safety reasons steel must be finished with layers guaranteeing sufficient fire resistance, such as sprayed light-weight concrete.

7.4 References and literature

[7.1] Gerstle, K. (1967). *Basic Structural Design.* McGraw-Hill Book Company, 405 pp.

[7.2] Beranek, L. (1971). *Noise and Vibration Control.* McGraw-Hill, New York.

[7.3] DBZ, Forschung und Praxis, Doppelboden-Anlagen, 76/7, p. 917 (in German).

[7.4] University of Illinois at Urbana-Champaign, Small Homes Council, Building Research Council, Council Notes F4.0, Wood Frame Floor Systems, 1978, 8 p.

[7.5] Prestatiegids voor gebouwen 5, Vloeren en trappen, IC-IB, 1980 (in Dutch).

[7.6] The Art of Construction, Architects Journal, 1981.

[7.7] Hens, H. (1982). *Bouwfysica 2, Warmte en Vocht, Praktische problemen en toepassingen.* ACCO Leuven (in Dutch).

[7.8] STS (1983). *Eengemaakte technische specificaties 23.* Houtbouw (in Dutch).

[7.9] Lutz, Jenisch, Klopfer, Freymuth, Krampf (1989). *Lehrbuch der Bauphysik.* B. G. Teubner, Stuttgart, 711 p. (in German).

[7.10] Cziesielski, E. (1990). *Lehrbuch der Hochbaukonstruktionen.* B. G. Teubner, Stuttgart, 627 p. (in German).

[7.11] Frick, Knöll, Neumann, Weinbrenner (1990). *Baukonstruktionslehre, Teil 1.* B. G. Teubner, Stuttgart, 588 p. (in German).

[7.12] Wiselius, S. I. (1996). *Houtvademecum.* Kluwer Techniek, 380 p. (in Dutch).

[7.13] Building Science Corporation (1998). *Builder's Guide, Hot-Dry & Mixed-Dry Climates.*

[7.14] Building Science Corporation (2000). *Builder's Guide, Hot-Humid Climates.*

[7.15] Beerepoot, M. (2002). *Energy regulations for new buildings, in search of harmonisation in the EU.* DUP Science.

[7.16] Leistner, P., Schröder, H., Richter, B. (2003). Gehgeräusche bei Massiv- und Holzbalkendecken. *Bauphysik 25,* Heft 4, S. 187–196 ((in German).

[7.17] Lang, J. (2004). Luft- und Trittschallschutz von Holzdecken und die Verbesserung des Trittschallschutzes duch Fussböden auf Holzdecken. *WKSB,* Heft 52, S. 7–14 (in German).

[7.18] Hens, H. (2005). *Toegepaste bouwfysica en prestatieanalyse 2/1a, bouwdelen.* ACCO, Leuven, 177 pp. (in Dutch).

[7.19] ASHRAE (2007). *Handbook of HVAC-applications, Chapter 43.* Atlanta.

[7.20] Georgescu, M. (2009). *Study of significant parameters for thermal insulation detailing at floor junctions with the external walls. Energy Efficiency and New Approaches* (Eds.: N. Bayazit, G. Manioglu, G. Oral and Z. Yilmaz). Istanbul Technical University, pp. 333–338.

[7.21] Glass, S., Charles, C., Curole, J, Voitier, M. (2010). *Moisture performance of insulated, raised wood-frame floors: a study of twelve houses in Southern Louisiana.* Proceedings of the Performance of Exterior Envelopes of Whole Building XI Conference, Florida (CD-ROM).

[7.22] Hens, H. (2010). *Prestatieanalyse van bouwdelen 1.* Uitgeverij ACCO, 327 pp. (in Dutch).

8 Outer wall requirements

8.1 In general

With the outer walls, building shape enters the picture. In general, envelopes include the opaque and transparent facade and roof parts plus the floors enclosing the conditioned volume. The main function of the envelope is to shield the indoors from the weather, be it that a well-designed envelope helps a lot in controlling thermal comfort, upgrading indoor air quality and increasing energy efficiency. In short, indoor conditions, outdoor conditions and their differences form the load the envelope has to tolerate.

The three chapters that follow focus on the opaque facade parts. Plenty of outer wall types find application. Massive, cavity and panelised walls are discussed here. In the first chapter of the second volume, wood-frame and metal-based types pass review, with lower front, curtain walls and double skin facades treated after the chapter on glazing and windows.

Structurally, outer walls are classified as load bearing and non-bearing. Load-bearing walls transmit their own weight and part of the deck load to the foundations, while adding stiffness against horizontal loads (wind, earthquakes, and obliqueness). They are usually heavyweight, with wood-frame as the exception. Non-bearing walls instead transmit own weight and wind load storey-wise to the load-bearing building structure. As these loads are usually quite low, lightweight designs are possible, with as extreme example the curtain wall. Non-bearing walls do not add stiffness to the building.

To summarize:

	Load-bearing	Non-bearing
Light-weight	(x)	x
Heavy-weight	x	x

8.2 Performance evaluation

8.2.1 Structural integrity

Neither allowed stress nor safety against collapse should be exceeded anywhere in an outer wall system, while maximum horizontal deflection must be limited to $1/500^{th}$ of storey height. With non-bearing outer wall systems, supports and suspensions must additionally guarantee stability against wind and earthquakes.

8.2.2 Building physics: heat, air, moisture

The requirements depend among others on the energy performance pursued:

- *Insulated buildings*
  A good thermal insulation is the only objective. It imposes requirements on the whole wall thermal transmittances. In some countries, the requirements depend on building compactness, i.e. the ratio between conditioned volume and enclosing envelope.
- *Energy efficient buildings*
  Combine good thermal insulation and correct ventilation with optimal use of solar and internal gains. Besides a low mean thermal transmittance, glass area, orientation and solar transmittance are important.
- *Low energy buildings*
  The objective is low primary energy consumption. For the low energy status of residential buildings, the annual primary energy for heating should not exceed 30 to 60 MJ per m^3 of protected volume, whereas passive measures must guarantee good summer comfort. Envelope related requirements are a mean thermal transmittance between 0.25 and 0.5 W/(m$^2 \cdot$ K) depending on building compactness and, air-tightness high enough to keep n_{50} below 3 ach or 1 ach when balanced ventilation with heat recovery is used (n_{50} is the ventilation rate at 50 Pa air pressure difference indoors/outdoors, measured with a blower door). Typical for outer walls are whole wall thermal transmittances around 0.2–0.25 W/(m$^2 \cdot$ K).
- *Passive buildings*
  Here the objective is very low primary energy consumption. Net energy demand for heating in residential buildings is limited to 18 MJ per m^3 of conditioned volume, whereas passive measures must guarantee good summer comfort. Envelope related requirements are a mean thermal transmittance between 0.15 and 0.3 W/(m$^2 \cdot$ K) depending on building compactness and, air-tightness high enough to keep n_{50} below 0.6 ach. For the outer walls, thermal transmittances of 0.1 W/(m$^2 \cdot$ K) are common despite the fact that such low values are beyond the life cycle cost optimum.
- *Zero and plus energy buildings*
  Zero energy means producing renewable energy so that non-renewable use is compensated on an annual basis. A positive energy balance means more renewable produced than the non- renewable annual consumption The outer wall requirements equal those for low energy or passive buildings.

8.2.2.1 Air tightness

Air tightness combines in- and exfiltration, wind washing in and behind the insulation, inside air washing in and in front of the insulation and, air looping around the insulation. 'Airtight' is an absolute requirement. As this is not reachable, an upper pivot is advanced. To prevent in- and exfiltration from causing problems, the mean air permeance coefficient of any outer wall may not exceed 10^{-5} kg/(m$^2 \cdot$ s $\cdot$ Pa^b), while concentrated leaks must be avoided. At the same time, enthalpy flow by wind washing, inside air washing and air looping may not increase heating season mean effective thermal transmittance by more than 10% compared to the conduction based whole wall value. For low energy, passive, net zero and plus zero energy, a 5% increase is the limit.

8.2.2.2 Thermal transmittance

Whole wall thermal transmittance is calculated as:

$$U = U_o + \frac{\sum(\psi_j L_j) + \sum \chi_k}{A_{deel}} \tag{8.1}$$

where U_o is the clear wall thermal transmittance, ψ_j the linear thermal transmittance of all linear thermal bridges, L_j their length and χ_k local thermal transmittance of all local thermal bridges present. Limit values (U_{max}) differ between countries, see Table 8.1.

Table 8.1: Limit values for the outer wall thermal transmittance.

Country	$U_{max} \leq$ W/(m²·K)
Belgium (Flanders)	0.40 (until 1/1/2012)
	0.32 (between 1/1/2012 and 1/1/2014)
	0.24 (from 1/1/2014 on)
Canada (Quebec)	0.28
Denmark	0.20
Sweden	0.1–0.2
Norway	0.2–0.3
Germany	0.35 (2008)
Finland	0.25 (2003)
England/Wales	0.35
Scotland	0.27
France	0.45 (2008)
The Netherlands	0.37 (2008)
Low energy buildings	0.2–0.3
Passive buildings	< 0.10–0.15

It is unclear if these maxima relate to the whole or clear wall thermal transmittance. Except when mentioned otherwise, all values in the chapters that follow assume an enthalpy flow of zero and a moisture response such that layers that must stay air-dry, stay. Passive buildings have the lowest pivot value, which, as was mentioned, is beyond the life cycle cost optimum. Low energy buildings target that optimum, equal to 0.2 to 0.25 W/(m²·K) depending on the insulation material used. That interval remains optimal even with external costs taken into account, at least in moderate climates.

Thermal bridge impact should stay below 0.1 U_o in insulated and energy efficient buildings and below 0.05 U_o in low energy buildings:

$$\frac{\sum(\psi_j L_j) + \sum \chi_k}{A} \leq 0.1 U_o \qquad \frac{\sum(\psi_j L_j) + \sum \chi_k}{A} \leq 0.05 U_o \tag{8.2}$$

The requirements for passive buildings are even more severe. Linear thermal transmittances should not exceed 0.01 W/(m·K), except at window reveals where 0.05 P/A is allowed, P being the window perimeter in m and A the window surface in m².

8.2.2.3 Transient response

Elementary check

For insulated and energy efficient buildings pivot dynamic thermal resistance is $1/U_{max}$ m²·K/W with U_{max} the legal (whole?) thermal transmittance limit value. Low energy buildings require a dynamic thermal resistance beyond 5 m²·K/W whereas passive buildings need 10 m²·K/W or more. Temperature damping beyond 15 and thermal admittances exceeding half the inside surface film coefficient ($h_i/2$ W/(m²·K)) are an advantage, but not a necessity.

Base check

Dynamic thermal resistance and admittance of the opaque outer walls must be such that for a given glass type, area, orientation and solar shading, for given infiltration rate, intentional ventilation, internal heat gain, admittances of the internal partitions and dynamic thermal resistance plus admittance of all other envelope parts enclosing the space, the number of excess temperature hours (WET-hours) should not exceed 100 annually. That number follows from the yearly sum of hourly mean thermal comfort weighting factors WF during the hours of building use:

$$|PMV| \leq 0.5 \quad WF = 0$$

$$|PMV| > 0.5 \quad WF = 0.47 + 0.22\,|PMV| + 1.3\,|PMV|^2 + 0.97\,|PMV|^3 - 0.39\,|PMV|^4$$

In the formulas, PMV is the hourly mean predicted mean vote. Calculations are done for the warm thermal reference year of the location considered.

The following simple method guarantees GTO ≤ 100 in a moderate climate without the need for a whole year simulation:

- Daily mean (operative) temperature indoors ($\bar{\theta}_o$) in a space during a representative hot summer day at the end of a heat wave at or below 28 °C
- Daily space damping obeying:

$$D_{\theta, space} \geq \frac{2.35 - 0.0224\,\bar{\theta}_o}{1 - 0.032\,\bar{\theta}_o}$$

The inside temperature $\bar{\theta}_o$ follows from a steady state heat balance taking into account all heat flows (transmission, infiltration, ventilation, solar gains, long wave losses to the sky, internal gains) and the sol-air temperature per envelope part. The daily zone damping stands for the harmonic temperature damping of a zone for a complex temperature outdoors with amplitude 1, assuming no infiltration, no intentional ventilation and neither solar nor internal gains.

8.2.2.4 Moisture tolerance

Water

- *Building moisture*
  Probability the outdoor wall turns air-dry without damage within an acceptable period (from one up to a couple of years) beyond 95%.
- *Rising damp*
  To be excluded by appropriate measures.
- *Wind driven rain*
  Wind driven rain mainly humidifies the windward side of a building with the highest intensities noted in the corners at the top. As a rule, a well-designed rain screen system should minimize penetration probability to less than 1%. Penetration happens each time rain wets the thermal insulation and/or the layers behind. Also, solar driven diffusion of rainwater buffered in capillary layers at the outside of the insulation can cause wetting.

Water vapour

- *Mould*
  The probability that mean relative humidity somewhere on the inside surface of an opaque outer wall part exceeds 80% on monthly basis (months set equal to a four weeks period) is below 5%. In moderate climates that condition is met for a temperature ratio passing 0.7.
- *Surface condensation*
  The probability that relative humidity somewhere on the inside surface of an opaque outer wall part will equal 100% at the design outside temperature is below 5%. In moderate climates this condition is met for a temperature ratio passing 0.7.
- *Interstitial condensation*
  The probability that moisture deposited will accumulate over the years somewhere in the assembly is below 1%. The probability to have yearly returning winter condensate beyond the damage threshold of the material(s) wetted is below 5%.

8.2.2.5 Thermal bridging

For the energy related requirements, see the paragraph on thermal transmittances. In addition, inside temperature ratios anywhere on a thermal bridge should not drop below 0.7:

$$f_{h_i} = \frac{\theta_{si,min} - \theta_e}{\theta_i - \theta_e} \geq 0.7 \tag{8.3}$$

If that condition is fulfilled, then, as just stated, probability to see mould developing or surface condensate deposited will stay below 5%.

8.2.3 Building physics: acoustics

For most outer wall systems, sound transmission loss of airborne noise produced outdoors is the most important performance to be met. Assessment in Europe is based on the EN-standards. These express sound transmission loss in dB(A). The value (R_w) is found by shifting the ISO reference curve against the measured or calculated sound transmission loss until both coincide on average. R_{500}, sound transmission loss at 500 Hz, read on the shifted reference and corrected

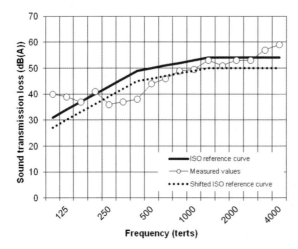

Figure 8.1: Determination of R_{500}, using the ISO reference curve.

for the specific frequency spectrum of the noise by traffic or other sources, then quantifies the wall's performance (Figure 8.1).

Considering the environment, the following requirements for the envelope, included the glazed surfaces, pertain for residential buildings:

Environment	Equivalent sound pressure level dB(A)	R_w dB(A)
Rural, sub-urban	$L_{aeq} \leq 55$	No requirement
Urban residential	$55 < L_{aeq} \leq 65$	$22 < R \leq 27$
Light industry, mixed commercial and residential	$65 < L_{aeq} \leq 75$	$27 < R \leq 32.5$
City centres, heavy industry, heavy traffic	$L_{aeq} > 75$	$32.5 < R \leq 37.5$

Also parallel paths via the outer walls should not drop sound transmission loss of party walls below the requirement imposed by the standards or by law. Between apartments and dwellings, sound transmission loss should equal 52 dB(A) or, if high quality is demanded, 60 dB(A). For buildings subjected to environmental legislation, the envelope must vice versa protect the neighbourhood from noise produced in the building. This is true for industrial premises, pubs, dance halls and others.

8.2.4 Durability

Besides a correct moisture tolerance, an additional drawback concerns hygrothermal loading ending in inacceptable cracking of outer wall layers during expected service life. Inacceptable means the cracks degrade facade architecture, compromise rain tightness and reduce air tightness. That risk should stay below 5%. Evaluation demands an analysis of crack initiation, based on a representative climate load in terms of temperature, solar radiation, night-time under-cooling, relative humidity, wind driven rain and frost.

8.2.5 Fire safety

Requirements consider fire class of the outside cladding, the insulation material and the inside finish, fire resistance of the outer walls and path length between window openings in successive storeys to avoid flame spread (Figure 8.2).

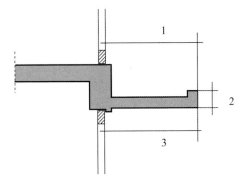

Figure 8.2: Path length $1 + 2 + 3 \geq 1$ m.

	Number of floors			
	1	2	> 2, highest floor < 22 m above grade	Highest floor > 22 m above grade
Material class	C	C	B	A
Fire resistance	–	–	–	90'
Path length between openings	–	–	1 meter	1 meter

8.2.6 Maintenance and economy

No general requirements, although life cycle cost analysis should be the reference.

8.3 References and literature

[8.1] Becker, R., Paciuk, M. (1996). *Application of the Performance Concept in Buildings.* Proceedings of the 3th CIB-ASTM-ISO-RILEM Intenational Symposium, Vol. 1 and 2.

[8.2] Hendricks, L., Hens, H. (2000). *Building Envelopes in a Holistic Perspective, Methodology.* Final Report IEA-ECBCS Annex 32 'Integral Building Envelope Performance Assessment', ACCO, Leuven.

9 Massive outer walls

9.1 Traditional masonry walls

9.1.1 In general

The traditional massive masonry wall was a brick and a half thick (Figure 9.1). It could be thicker if needed for structural reasons, but if thinner the nasty result was degraded moisture tolerance. In spite of this, many 19-century working class houses had one-brick outer walls.

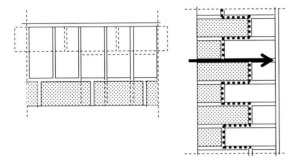

Figure 9.1. Traditional brick and a half thick wall. The dotted line shows the continuous vertical mortar/brick interface acting as rain barrier in the wall, the black arrow signals the interruption at each bed joint.

9.1.2 Performance evaluation

Given the fact that mechanical integrity, included buckling, hardly caused problems with traditional masonry walls, the discussion focuses on building physics and durability.

9.1.2.1 Building physics: heat, air, moisture

Air tightness

For a brick and a half thick wall, plastered at the inside, the air permeance coefficient drops below 10^{-5} kg/(m² · s · Pa^b), i.e. stays below the threshold for acceptability. Without plaster at the inside, the probability to see the air permeance coefficient increase to values substantially above that threshold is high.

Thermal transmittance

Air-dry, the clear wall thermal transmittance of a brick and a half thick outer wall reaches 1.9 W/(m² · K), far too high with respect to the values imposed today. Moreover, the value fluctuates with moisture content. At the rain side, it increases to 2.3 W/(m² · K) in winter. That of course is an additional disadvantage, which rules out applicability.

Transient response

The only control possible at outer wall level is to check if the dynamic thermal resistance exceeds $1/U_{max}$ with U_{max} the legally imposed thermal transmittance upper limit, if temperature damping exceeds 15 and if admittance goes beyond $h_i/2$. For a brick and a half thick wall these properties calculate as:

Dynamic thermal resistance		Temperature damping		Admittance	
D_q $m^2 \cdot K/W$	Phase angle h	D_θ –	Phase angle h	Ad $W/(m^2 \cdot K)$	Phase angle h
1.2	8	5.7	9	4.7	1

Only the admittance meets the requirement. The dynamic thermal resistance remains far below the desired value, while temperature damping does not even near the pivot 15.

Moisture tolerance

Building moisture

Thanks to the low critical moisture content of bricks ($w_{cr} \leq 100$ kg/m³) and the low equivalent vapour resistance factor of masonry ($5 < \mu_{eq} < 10$), a brick and a half thick wall dries quite fast. Of course, painting should not be done until the wall is air-dry.

Rising damp

Rising damp should not be a problem on condition the brick and a half thick wall gets a damp proof layer inserted above grade and above all locations where rain run-off may collect and be absorbed by the wall (Figure 9.2).

Older buildings often got a damp proof course of very dense bricks up to a few layers above grade. But rising damp via the bed and head joints remained a complaint. Solutions in the case that rising damp still causes problems include: (1) inserting a damp proof layer.

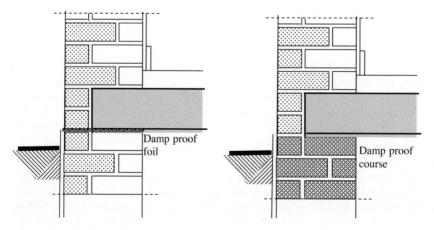

Figure 9.2. Damp proofing.

9.1 Traditional masonry walls

If sawing the brick wall is impossible, that option requires a lot of chopping and breaking, (2) injecting the wall above grade or locations where rain run-off collects with a water repellent or pore filling product. Drying pipes or the application of electro kinesis to reverse capillary action have no effect.

Very important is to test beforehand whether hydrating salts are the cause of dampness. If so, neither injection nor a damp proof layer will prevent the wall from staying damp. The only solutions left in such cases are demolition, applying a restoration plaster or hiding the masonry at the inside behind a vapour-tight facing wall.

Wind-driven rain

Three facts minimize the probability of rain penetration across a brick and a half thick wall: a high buffer capacity (up to 70 kg water per m²), the usually high capillary absorption coefficient of bricks and mortar and a mortar interface dividing the wall in two halves, one acting as a water buffer and the other staying dry. The high capillary absorption coefficient delays the moment run-off starts during wind driven rain events. That limits water load on delicate details such as the window/masonry rebates. At the same time, the vertical mortar interface figures as additional liquid water resistance. Of course, that interface sees its integrity interrupted by all bed joints. These form easy moisture transfer paths when the mortar is capillary.

Mould

Table 9.1 gives the temperature ratios in the edge between two brick and a half thick outer walls ($h_i = 4$ W/(m²·K)), in the zone just above the skirting board ($h_i = 3$ W/(m²·K)) and behind cupboards against the outer wall ($h_i = 2$ W/(m²·K)).

Table 9.1. One and a half stone thick brick masonry outer wall.

Location	f_{hi}
Edge between two outer walls	0.64
Zone just above skirting board	0.57
Behind cupboards against the outer wall	0.48

Particularly the temperature ratio behind cupboards limits the allowable indoor vapour pressure excess to avoid mould on the inside surface. For a cool climate January, the mean excess should not exceed 205 Pa, a value so low that exceeding probability increases mould risk to 17%, quite higher than the 5% allowed.

Surface condensation

As for mould, also here the temperature ratio behind cupboards defines whether surface condensation will occur or not.

Interstitial condensation

Interstitial condensation is not a problem on condition the brick and a half thick wall has an air permeance coefficient around 10^{-5} kg/(m²·s·Pa^b). For that to be true, an inside plaster finish is mandatory. In case this is left out for architectural reasons, air-tightness drops considerably. High indoor vapour pressure excesses may then cause condensation in the wall.

Thermal bridging

A brick and a half thick wall acts as one large thermal bridge, see the temperature ratios of Table 9.1.

9.1.2.2 Building physics: acoustics

A brick and a half thick wall, plastered at the inside, stands for a sound transmission loss of 59 dB, i.e. truly well performing. In fact, the windows define the final sound insulation quality of an envelope, not the massive walls.

9.1.2.3 Durability

Thanks to the deformability of masonry, especially when brick-laid with lime or bastard mortar, the hygrothermal load only exceptionally creates problems.

9.1.2.4 Fire safety

A brick and a half thick wall, plastered at the inside, has a fire resistance exceeding 120′, which is largely sufficient.

9.1.3 Conclusion

The performance evaluation shows that above all deficient thermal quality has discredited the brick and a half thick outer wall. Upgrading may be done by:

1. Constructing the wall in lightweight blocks.
2. Insulating the wall at the inside
3. Insulating the wall at the outside

Option 1 can be applied in new construction, though the thermal transmittances imposed actually touch values no longer within reach of such walls. Option 2 and 3 are applicable in new construction and in retrofit. Still, some designers presume the three are qualitatively equivalent. That opinion is assessed in the paragraphs that follow.

9.2 Massive light-weight walls

9.2.1 In general

Widely used as lightweight blocks are cellular concrete and extra light fast bricks. Cellular concrete has a density ranging from 450 to 600 kg/m^3. That low density allows manufacturers to offer large format blocks, 30 × 30 × 60 cm. These are glued with thin-bed mortar to form half-block walls. The market also offers 60 cm high units (Figure 9.3).

Extra light fast bricks have a density not dropping below 780 kg/m^3. That higher weight obliges manufacturers to produce smaller blocks compared to cellular concrete. But also here, gluing with thin-bed mortars becomes popular (Figure 9.3). For both block types, head joints are replaced by groove and tongue locking.

9.2 Massive light-weight walls

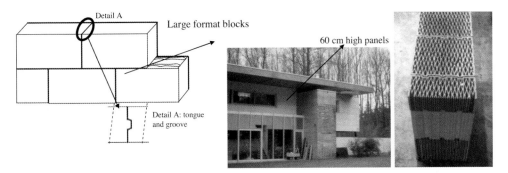

Figure 9.3. Massive walls in cellular concrete and in extra light fast bricks.

9.2.2 Performance evaluation

9.2.2.1 Structural integrity

The rules in force for massive load-bearing outer walls also apply to lightweight blocks: stiffening against wind, limiting buckling length, etc. Of course low compression strength and lower modulus of elasticity restricts the load-bearing capacity. With reinforced concrete ring beams cast under each floor deck to a maximum of 3 storeys. These ring beams redistribute concentrated loads alongside window and door bays over a greater wall length. Above bays that do not touch the ring beam, horizontal joints get trussed stainless steel reinforcement. (Figure 9.4).

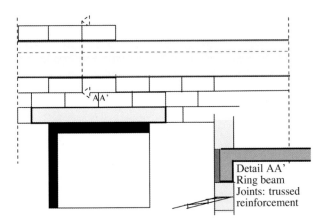

Figure 9.4. Ring beam under each floor deck redistributing loads. Above bays not touching the beam, joint reinforcement is used.

9.2.2.2 Building physics: heat, air, moisture

Air-tightness

In case the blocks are carefully glued and the half-brick walls get a plaster rendering at the inside and a stucco finish at the outside, the mean air permeance coefficient amply drops below 10^{-5} kg/(m² · s · Pa^b).

Thermal transmittance

For air-dry cellular concrete with a density of 450 kg/m³ and thermal conductivity reaching 0.14 W/(m · K) when glued and 0.18 W/(m · K) when brick-laid or, for air-dry light weight fast bricks with density 780 kg/m³ and thermal conductivity 0.2 W/(m · K), a clear wall thermal transmittance (U_o) 0.6 to 0.1 W/(m² · K) requires wall thicknesses equal to:

Material	Wall thickness in cm			
$U_o =$	0.6 W/(m² · K)	0.4 W/(m² · K)	0.2 W/(m² · K)	0.1 W/(m² · K)
Cellular concrete ($\lambda = 0.14$ W/(m · K) / $\lambda = 0.18$ W/(m · K))	21/27	33/42	68/87	138/177
Light weight fast bricks	30	47	97	197

If a thickness of 29 to 34 cm is economically feasible and in line with tradition, clear wall thermal transmittances below 0.4 W/(m² · K) rapidly lead to uneconomical thicknesses that are also structurally problematic. Moreover, to keep the value weather independent, a rain repelling outside finish is needed. That can be stucco or a siding.

Thermal bridging is mainly found at floor supports, window bays and door bays. Figure 9.5 shows the worst possible and two better ways of detailing these.

Accompanying linear thermal transmittances are listed in Table 9.2.

Table 9.2. Cellular concrete, glued ($\lambda = 0.14$ W/(m · K), thickness 30 cm, $U_o = 0.42$ W/(m² · K)), details, linear thermal transmittances (ψ) (see Figure 9.5).

Detail	Construction	ψ W/(m · K)
Floor support	Traditional	0.48
	Upgraded	0.19
	Best	0.11
Lintel (30 cm high)	Traditional	1.09
	Upgraded	0.53
	Best	0.36
Window reveal	Window 7 cm behind frontal surface	0.01

9.2 Massive light-weight walls

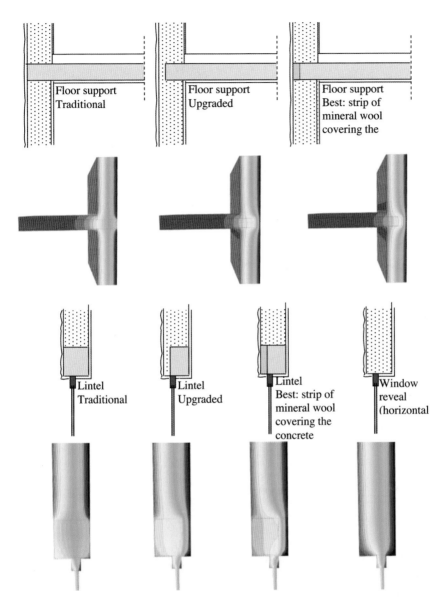

Figure 9.5. 30 cm thick cellular concrete wall: floor support, lintel and window reveal (The isotherms are reversed).

Transient response

Again, the only control possible is checking if the dynamic thermal resistance exceeds $1/U_{max}$ with U_{max} the legally imposed thermal transmittance, if temperature damping exceeds 15 and if the admittance goes beyond $h_i/2$. For a cellular concrete and lightweight fast brick wall, both with a thickness of 30 cm, the three values are (to repeat, they relate to a daily variation in the outdoor temperature):

	Dynamic thermal resistance		Temperature damping		Thermal admittance	
	m²·K/W	Phase, h	–	Phase, h	W/(m²·K)	Phase, h
Cellular concrete ($\lambda = 0.18$ W/(m·K))	5.2	9.6	11.3	12.0	2.2	2.4
Light weight fast bricks	6.5	11.0	16.4	13.3	2.5	2.3

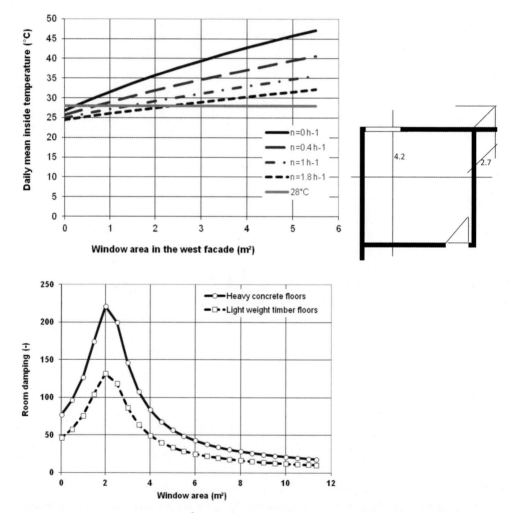

Figure 9.6. Room 4.2 × 4.2 × 2.7 m³, outer and inner walls 30 cm cellular concrete, variable window area in the west facade. Daily mean inside temperature at the end of a heat wave and room damping, both as function of window area.

9.2 Massive light-weight walls

While the dynamic thermal resistance exceeds the requirement for low energy buildings and temperature damping nears and reaches the pivot of 15, the admittance hardly exceeds half the value requested, which is 3.9 W/(m² · K) (= $h_i/2$). Both walls consequently figure as thermally inert but quite inactive as a heat storage medium. The reason is the combination of moderate density with low thermal conductivity. Because of that, summer comfort in residential buildings constructed with cellular concrete or lightweight fast brick walls may be disappointing without active cooling or any additional passive measure to moderate solar gains.

As an example, Figure 9.6 shows the daily mean inside temperature after a series of hot days in a corner room, floor area 4.2 × 4.2 m², height 2.7 m, of an apartment at an intermediate floor as a function of window area in the west facade. The outer and inner walls consist of 30 cm cellular concrete whereas the window frame contains x m² of low-e, argon filled double glass. Ventilation figures as a parameter.

With a window area below 2 m², which means 1.4 m² of glass and a window/floor ratio less than 1/8, daily mean inside temperature firstly stays below 28 °C for ventilation rates above 1.8 ach. Reaching the room damping needed – 16.7 – instead is not a problem, even with a lightweight timber floor. Or, building with lightweight 30 cm thick outer walls may easily result in summer discomfort. Anyhow, contrary to the thermal transmittance, transient thermal properties at outer wall level on their own are compelling performance indicators.

Moisture tolerance

Rising damp

Rising damp is easily avoided by inserting a damp proof layer in all outer and inner walls just above grade and above all location where rain run-off collects.

Building moisture

Building moisture is a potential nuisance with cellular concrete. Due to autoclaving, fresh blocks contain up to 200 kg moisture per m³. After construction, that wetness slowly dries until hygroscopic equilibrium, some 20 kg per m³. The rate first depends on relative humidity indoors. The higher it is, the lower that rate. Drying time further increases quadratically with wall thickness: $t_d \geq \beta \, d^2$ with β approximately equal to 5 · 10⁴ day/m² (second drying period). If drying proceeds at both sides, then d stands for half the thickness. In a moderate climate, it may take more than 12 years for a 30 cm thick outer cellular concrete wall with vapour-tight outside finish to reach hygroscopic equilibrium in an indoor climate class 2. If the same wall can dry to inside and outside, that period shrinks to 3 years. Or, the diffusion resistances of both the inner and outer finish are most influential for the drying rate. If an outer wall gets a vapour retarding finish at both sides, for example by using vapour retarding paints, drying proceeds extremely slowly, as Table 9.3 illustrates.

Clearly, especially the diffusion thickness (μd) of the outside finish determines drying times. Also, a more severe indoor climate slows drying. In indoor climate class 5 buildings complete drying becomes impossible once the outside finish has a diffusion thickness beyond 0.5 m.

The drawbacks of too slow drying are evident. As Figure 9.7 underlines, clear wall thermal transmittance surpasses the air-dry value for a too long period, resulting in more end energy consumed for heating. Cellular concrete walls, painted too early, may blister. Just finished new dwellings in cellular concrete could also suffer from high relative humidity in the first years, with mould as one of the consequences.

Table 9.3. 30 cm thick cellular concrete outer wall, drying in the cool climate of Uccle, Belgium (ICC 1 to 3, indoors 17 °C in January, 23 °C in July, ICC 5 stands for a natatorium).

Indoor climate class	μd inside finish (m)	Drying time till hygroscopic equilibrium in years		
		μd outside stucco (m)		
		0.15	0.5	1.0
1	0.025	5.4	9.5	13.5
2		6.5	12.5	> 20
3		7.5	15.5	> 20
5		10.5		
1	0.5	6.6	11.6	Complete drying impossible
2		7.6	14.6	
3		8.4	17.5	
5		8.4	> 20	

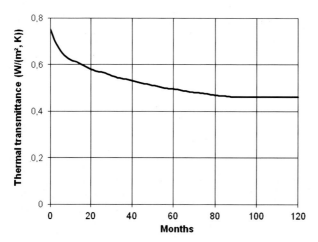

Figure 9.7. 30 cm thick cellular concrete outer wall. Clear wall thermal transmittance during drying (Indoor climate class 2, diffusion thicknesses: inside finish 0.025 m, outside finish 0.15 m).

The best way to shorten drying time is by using stuccoes with vapour diffusion thickness below 0.25 m when the relative humidity indoors is low and below 0.15 m when the relative humidity indoors is high, in combination with a vapour permeable inside finish (neither oil paint nor paints with great masking power). Also, ventilation between 0.5 and 1 ach during the first years of occupation is mandatory.

For lightweight fast brick outer walls, building moisture is not an issue. The outside stucco diffusion thickness requirement is therefore less severe, although 2 m is a preferred maximum.

Wind driven rain

Cellular concrete walls without outside finish suck rainwater quite slowly. Capillary water absorption coefficient (A) in fact does not exceed 0.047 kg/(m² · s$^{0.5}$)), while critical (w_{cr}) and capillary moisture content (w_c) score high, ≈ 180 kg/m³ and ≈ 350 kg/m³ respectively.

9.2 Massive light-weight walls

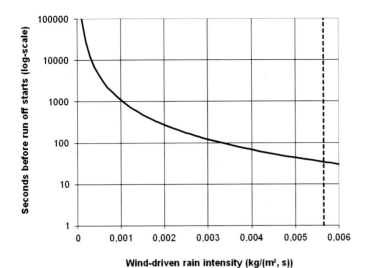

Figure 9.8. 30 cm thick cellular concrete outer wall, time in seconds between the start of a wind-driven rain event and run-off. The vertical line stands for the maximum wind-driven rain intensity measured at Uccle, Belgium.

When hit by wind driven rain, the three properties bring a quick reversal from buffering to run-off, as Equation (9.1) and Figure 9.8 underscore:

$$t_r = 0.62 \frac{A^2}{g_{ws}^2} \tag{9.1}$$

g_{ws} in that formula is wind-driven rain intensity in kg/(m² · s).

Moreover, the moisture front in the material moves so slowly that weeks of wind driven-rain are needed before the inside surface becomes wet, see Equation (9.2) and Figure 9.9:

$$t_2 = \frac{t_r}{2} + \left[\frac{d^2 (w_c + w_{cr})^2}{4 A^2} \right] \tag{9.2}$$

Even without stucco finish, the outside surface of a cellular concrete outer wall mainly functions as a drainage plane with buffering as a second order mechanism – see the moisture profile during the heating season of Figure 9.10. The profile $w_k(x)$ represents the maximum moisture content at each location in the wall, the profile $w_1(x)$ the moisture content exceeded 5% of the time and the profile $w_2(x)$ the moisture content exceeded half the time. At no point did the front penetrate more than 9 cm into the wall.

Nevertheless, finishing the wall with stucco or a siding is preferred. Run-off in fact leads to high rain loads on the joints between blocks and the joints between cellular concrete and other envelope parts. Water for example only needs 75 seconds to penetrate a shrinkage crack between two 30 cm thick blocks. Moreover, winter mean clear wall thermal transmittance increases substantially at the rain side, for a 20 cm thick cellular concrete outer wall from 0.77 to 1.31 W/(m² · K), i.e. +77%, for a 30 cm thick cellular concrete outer wall from 0.55 to 0.87 W/(m² · K), i.e. +58%. A capillary wet cellular concrete is also hardly frost resisting.

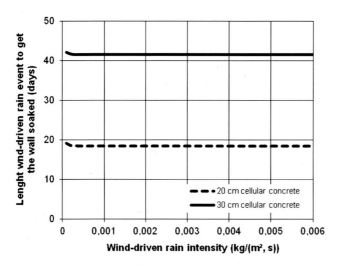

Figure 9.9. 30 cm thick cellular concrete wall, time in days wind-driven rain events must last to soak the whole wall.

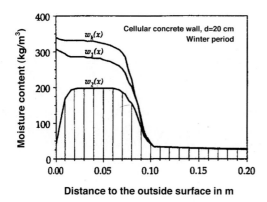

Figure 9.10. Moisture profiles during the heating season in a cellular concrete outer wall without outside finish (w_k: characteristic moisture profile, w_1: most common moisture profile, w_2: quasi-permanent moisture profile).

For an outside stucco to fulfil its role as rain protection, the capillary water absorption coefficient should not exceed $5 \cdot 10^{-5}$ kg/(m² · s$^{0.5}$).

For outer walls in lightweight fast bricks, an outside rain protecting finish is even mandatory. Not only is the likelihood to get soakage much higher than for cellular concrete, also the clear wall thermal transmittance increases more when wetted by wind-driven rain. Stucco requirements however are less severe. For a diffusion thickness of 2 m, capillary water absorption coefficient may touch 0.0083 kg/(m² · s$^{0.5}$). Their product anyhow must stay below 0.0033 kg/(m · s$^{0.5}$).

9.2 Massive light-weight walls

Mould and surface condensation

With a clear wall thermal transmittance below 0.6 W/(m^2·K) risk of mould developing or surface condensate depositing drops below 5%. Of course, thermal bridges with temperature ratios below 0.7 must be avoided. And, as said, preventing mould problems in cellular concrete buildings requires good ventilation the first years after occupation.

Interstitial condensation

On first sight and according to a Glaser calculation, avoiding problematic interstitial moisture deposit in lightweight massive walls is simple: use inside finishes that limit vapour diffusion to a level dictated by the outside finish diffusion thickness. Or, any type of outside stucco or cladding is applicable on condition the right inside finish is chosen.

However, for cellular concrete, this does not hold. Building moisture must dry out. For that to happen, a vapour permeable outside stucco or siding is a necessity. Moreover once air-dry, so-called interstitial condensation in indoor climate class 1 to 3, becomes a ripple in hygroscopic moisture content, somewhat higher after winter and lower after summer. Even the accumulation of condensate in indoor climate class 4 and 5 is hardly harmful. In fact, the moisture front starting behind the outside finish ultimately stabilizes with as moisture content in the wet zone the critical value. The only meaningful requirement therefore is that the clear wall thermal transmittance should not increase with 10% or more. That gives the following design rule for a 30 cm thick outer wall in a moderate climate with an annual mean dry bulb outside temperature of ≈ 10 °C:

$$\theta_c = 7.1 + 0.17\,\theta_{im} + 2.9\,a_K$$

$$(\mu d)_e \leq 0.25 \text{ m} \qquad (\mu d)_i \geq (\mu d)_e \frac{p_{im} - 298 - 92.4\,\theta_c}{92.4\,\theta_c - 11\,\theta_i - 710} - 1.27 \qquad (9.3)$$

θ_c is the annual mean temperature at the moisture front, a_K the short wave absorption factor of the outside surface, $(\mu d)_e$ the equivalent diffusion thickness of the outside finish, $(\mu d)_i$ the equivalent diffusion thickness of the inside finish, p_{im} the annual mean vapour pressure indoors and $\theta_{i,m}$ the annual mean temperature indoors ($15 \leq \theta_{i,m} \leq 30$ °C). If $(\mu d)_i$ turns negative, any inside finish may be used. For each indoor climate class, one gets:

Indoor climate class	Diffusion thickness inside finish $(\mu d)_i$ in m
1	No requirements
2	No requirements
3	No requirements
4, 5	$(\mu d)_e \leq 0.15$ m, $(\mu d)_i \geq 0.4$ m

For lightweight fast brick outer walls, the same approach applies. For indoor climate class 3 there are no requirements, in indoor climate class 4 and 5 they are: $(\mu d)_I \geq 1.3\,(\mu d)_e$ with $(\mu d)_e \leq 2$ m.

Thermal bridging

Lintels, window reveals, sills, floor supports, foundation/floor on grade nodes and roof parapets are critical. Some are catalogued in Figure 9.5. That figure also shows a preferred solution:

Table 9.4. Cellular concrete outer wall, 30 cm thick, joints glued ($\lambda_d = 0.14$ W/(m·K)), temperature ratio for the details shown in Figure 8.8.

$U_o = 0.42$ W/(m²·K) Detail	Construction	f_{hi} —
Floor support	Traditional	0.70
	Upgraded	0.81
	Best	0.83
Lintel (30 cm high)	Traditional	0.48
	Upgraded	0.65
	Best	0.71
Window reveal	Window 7 cm behind frontal surface	0.69

insulating all exposed heavy concrete parts with strips of hard mineral wool. In case the wall's stucco finish is not reinforced, embed a glass fabric over these strips. Table 9.4 lists temperature ratios. For the floor support, the value given assumes a cupboard standing against the facade wall. The difference between traditional and preferred (called best) is obvious. All best solution values approximate or exceed 0.7, limiting mould risk to less than 5%.

For lightweight fast bricks, analogous numbers are found.

9.2.2.3 Building physics: acoustics

Compared to massive masonry, lightweight outer walls lose quite some sound insulating quality. A 30 cm thick cellular concrete wall has a sound transmission loss of 45 dB, whereas a 30 cm thick masonry wall gave 59 dB. With lightweight fast bricks, the result depends on the perforation pattern, as Figure 9.11 shows. Still, the sound insulating quality remains high enough for the windows to define the result.

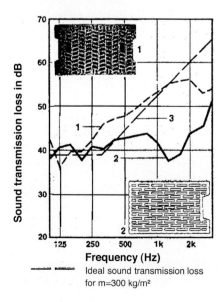

Figure 9.11. Sound transmission loss of light weight fast brick outer walls depending on perforation pattern (1 uncrossed, 2 crossed).

9.2 Massive light-weight walls

9.2.2.4 Durability

As long as a cellular concrete wall contains building moisture, hygric strain defines stress response. Once air-dry, thermal loading takes over. Typical for lightweight massive walls is that on an annual basis mean temperature and temperature difference across the wall fluctuate heavily, see Figure 9.12.

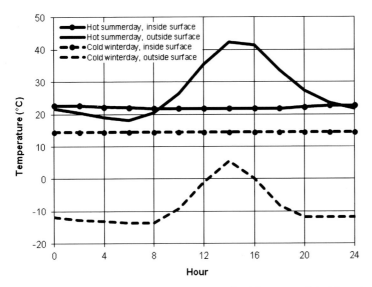

Figure 9.12. 30 cm thick cellular concrete wall, white stuccoed, inside and outside surface temperatures during a hot summer and cold winter day.

That causes length changes and deflection. In case the outside stucco is stiffer and has a higher thermal expansion coefficient than the blocks, the neutral axis in the wall moves in the direction of the stucco. Even changes in mean temperature then cause bending, to the outside for a temperature increase and to the inside for a temperature drop. For a 30 cm thick cellular concrete wall the stresses induced in the stucco that way may exceed its tensile strength with crack initiation and crack growth as a result. Once the cracks touch the blocks, the stucco loses rain screen ability.

A good outside stucco must thus have (1) a modulus of elasticity (E) equal or lower than the lightweight masonry, (2) a high tensile strength (σ) and (3) show a hygric (ε_h) and thermal expansion (α) hardly different from the cellular concrete or the lightweight fast bricks. These three conditions combined, give the ratios $E\alpha/\sigma$ and $E\varepsilon_h/\sigma$. The lower both are, the more suitable the stucco. Of course, in reality, the two cannot be manipulated freely. A lower modulus of elasticity means on average a lower tensile strength, which is why additional measures, such as reinforcing the stucco with glass fibre fabric are necessary.

A detailed study of stuccos for cellular concrete walls resulted in the following performance requirements:

- Thermal expansion coefficient between $7.2 \cdot 10^{-6}$ and $10.4 \cdot 10^{-6}$ K^{-1}
- Product of modulus of elasticity and thermal expansion coefficient below 0.017 MPa/K

- Hygric shrinkage during drying from capillary moist to air-dry at 33% relative humidity between $5 \cdot 10^{-4}$ and $7.2 \cdot 10^{-4}$ m/m
- Product of modulus of elasticity and hygric shrinkage below 1 MPa
- Relation between modulus of elasticity (E), thermal expansion coefficient (α) and tensile failing strain (ε_{fr}):

$$E \leq 1650 \text{ MPa}: \quad E^2 \alpha / \varepsilon_{fr} \leq 1120 \qquad E > 1650 \text{ MPa}: \quad E \alpha / \varepsilon_{fr} \leq 0.68 \qquad (9.4)$$

- Adhesion above 11 ε_{fr} (in MPa)

Table 9.5 gives measured data for six stuccos. Only the lightweight 1 and 2 fulfil all requirements (greyed). Both are fibre reinforced, water repelling mineral stuccos.

Table 9.5. Properties of some lightweight (L) and mineral stuccos (M).

Stucco	α (· 10^{-6}) K^{-1}	E MPa	ε_{fr} m/m	$E\alpha$ MPa/K	$E^2\alpha/\varepsilon$ MPa^2/K	$E\alpha/\varepsilon$ MPa/K
Light-weight 1	9.3	340	0.001765	0.0032	609	
Light-weight 2	10.3	610	0.004262	0.0063	899	
Light-weight 3	10.4	740	0.002568	0.0077	2218	
Mineral 2	4.7	4460	0.000516	0.021		40.6
Mineral 2	10.7	2540	0.000630	0.027		43.1
Mineral 2	12.5	3910	0.0001023	0.049		47.8

9.2.2.5 Fire safety

The fire resistance of massive walls built with lightweight blocks of fire reaction class A1 is excellent. An overall fire resistance of 120′ does not pose any problem. One of the reasons is the excellent thermal resistance, which limits temperature increase at the non-fired surface, whereas thermal expansion of the blocks causes compression rather than initiating cracking.

9.2.2.6 Maintenance

Maintenance concerns the stucco. The reasons for degradation are crack initiation and growth, fouling where there is no run-off or run-off from dirty horizontal or weakly sloped surfaces (for example sills) and algae growth on permanently humid surfaces. In the first years, hosing off suffices. After 10 to 12 years, repainting may be required.

9.2.3 Design and execution

9.2.3.1 In general

Lightweight massive outer walls have limitations. A 'normal' wall thickness of 30 cm gives a clear wall thermal transmittance 0.45 to 0.6 W/(m$^2 \cdot$ K), which excludes their usage in low energy buildings, where whole wall thermal transmittances of 0.2 W/(m$^2 \cdot$ K) or below is the pivot. Other conclusions out of the performance evaluation are:

9.2 Massive light-weight walls

- Contrary to what manufacturers claim, massive walls of lightweight blocks are no magic solution in terms of transient thermal response. On the contrary, admittance is so low they react more like light-weight wall systems
- For rain tightness reasons an outside stucco or siding is recommended. Only for cellular concrete walls applied in industrial premises can a finish be omitted
- The stucco used has to be vapour permeable, hardly capillary and deformable (low modulus of elasticity, sufficient ductility)
- Exposed concrete parts should be finished with a stuccoed or sided mineral wool strip at the outside.
- Take care with 'integral cellular concrete constructions'. The material performs adequately under compression but less when bent. Prefab cellular concrete floor slabs are also more deformable and creep sensitive than cast or prefabricated concrete ones

9.2.3.2 Specific

Large format blocks shorten building time. Thin bed mortars for the bed joints and groove and tongue locking of head joints further simplify brick laying. The horizontality of the first row is critical. That is taken care of by brick laying it along the string using small format blocks (Figure 9.13).

Large format blocks of course only make sense if the design is modular, for example on a 30 cm base. If not, much time and material is lost in abating the blocks, which makes economics and sustainability disputable. Lightweight blocks facilitate channelling of pipes, but pipes concentrated in shafts that allow control and repair are to be preferred.

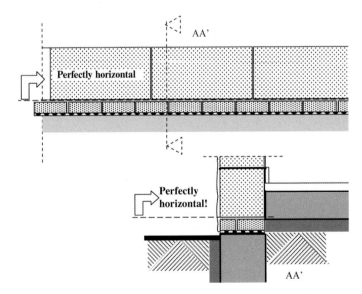

Figure 9.13. Thin bed mortared large format block masonry, first row (with perimeter and floor insulation indicated).

9.3 Massive walls with inside insulation

9.3.1 In general

Distinction is made between two solutions: stud based and sandwiches. Stud-based starts with mounting timber studs against the inside surface of the massive wall, followed by filling the bays with mineral wool, stapling, when needed, a vapour retarding foil against the studs and lining up the whole with gypsum board or gypsum plaster sprayed on metal cloth (Figure 9.14). Sandwiches consist of storey-high 'thermal insulation (MW, PUR, EPS, XPS)/gypsum lining' boards, glued against the inside surface (Figure 9.14).

A recent method is the use of capillary insulation materials, a typical example being porous calcium silicate boards. The intent is to avoid wetting of the massive wall by interstitial condensation and to exclude the need for a vapour retarder. The capillary boards are glued against the wall with special mortar and finished with spray gypsum.

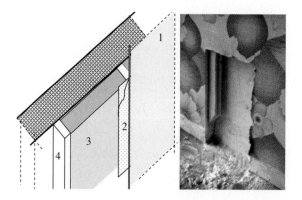

1=gypsum board, 2=vapour retarder, 3=thermal insulation, 4=timber-frame

Figure 9.14. Inside insulation. Left: stud based, right: sandwich.

9.3.2 Performance evaluation

9.3.2.1 Structural integrity

In general, structural integrity hardly poses any problem. Of course, the insulation boards must bear their own weight without yielding. The situation changes in case the massive wall is not airtight and the insulation is. In such case, the insulation system has to withstand wind load, demanding a bond to the wall strong enough to resist even high wind pressure without structural damage.

9.3.2.2 Building physics: heat, air, moisture

Air tightness

The prime requirement for good air-tightness is an air leakage free base wall. Otherwise the inside insulation becomes a wind barrier and loads accordingly. A concrete wall has an air permeance close to zero. For a well-pointed one-brick masonry wall that value reaches

9.3 Massive walls with inside insulation

$1.5 \cdot 10^{-5} \Delta P_a^{-0.19}$ s/m. Masonry in porous concrete blocks see that value increase by a factor of 25, up to $4 \cdot 10^{-4} \Delta P_a^{-0.25}$ s/m. An inside plaster or outside stucco drops that number to $10^{-5} \Delta P_a^{-0.2}$ s/m, the performance requirement set for outer walls. Also, the insulation system is preferentially airtight. For the timber-framed type, the correctly jointed gypsum board lining, with an air permeance below $3 \cdot 10^{-5} \Delta P_a^{-0.19}$ s/m, or the spray gypsum plaster, both with vapour retarder if necessary, should guarantee air-tightness.

However, open joints across the insulation at skirting and ceiling level and a continuous air layer between insulation and wall creates winter upside down indoor air washing in moderate and cold climates. How significant washing is and how detrimental for insulation performance and moisture tolerance depends on the width of both the air layer and the open joints – see Figure 9.15.

Thermal transmittance

Wall airtight, no indoor air washing

In case the wall is airtight and indoor air washing non-existent, then the clear wall thermal transmittance (U_o) mainly depends on the thickness and nature of the insulation material. Thermal resistance of the basic wall in fact hardly plays a role. With values between 0.6 and 0.1 W/(m²·K), a one-brick wall requires insulation thicknesses of:

U_o- value W/(m²·K)	Insulation thickness cm				
	MW	PUR	EPS	XPS	CG
0.6	4	3	4	3	5
0.4	8	5	8	6	9
0.2	17	12	18	13	20
0.1	37	26	38	28	42

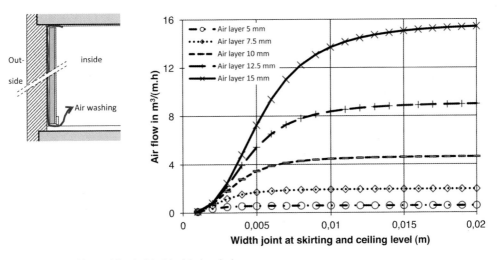

Figure 9.15. Air washing behind inside insulation.

For a value of 0.6 W/(m² · K) insulation thickness remains very moderate, resulting in an overall wall thickness below 30 cm. A value 0.4 W/(m² · K) increases that overall thickness to some 30 cm. At values of 0.1 W/(m² · K), however, the insulation and total thickness explode to values between 48 and 64 cm. Compared to a "normal' wall of 30 cm, the last number gives an inside floor area loss of 0.34 m² per meter run of facade wall, which must be accounted for when looking to the life cycle costs!

Yet, with inside insulation, the clear wall thermal transmittance has hardly any significance. The solution is caught between two non-solvable thermal bridges: floor deck supports and junctions between partition walls and outer walls. Figure 9.16 gives the isotherms for a concrete structure, composed of a 20 cm thick outer wall with 6 cm insulation inside (λ = 0.04 W/(m² · K)) and a 20 cm thick concrete floor with screed touching the outer wall. Table 9.6 lists the whole wall thermal transmittance for that outer wall, now 4.8 × 3 m² large and enclosed by two partition walls and two floors. The table is self-explaining.

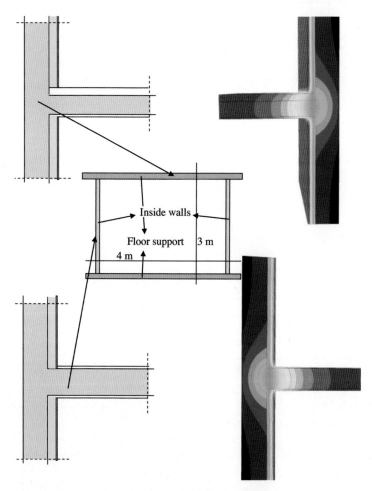

Figure 9.16. Concrete structure insulated at the inside, floor support and inner/outer wall junction (isotherms going from 0 to 20 °C).

9.3 Massive walls with inside insulation

Table 9.6. Concrete structure, insulated at the inside: whole wall thermal transmittance.

Insulation thickness (λ = 0.04 W/(m·K))	Clear wall U-value	Linear thermal transmittance		Whole wall U-value	Increase in %
		Partition wall junction ψ	Floor support ψ		
m	W/(m²·K)	W/(m·K)	W/(m·K)	W/(m²·K)	
0.06	0.55	0.87	0.84	1.01	84
0.08	0.43	0.86	0.83	0.89	107
0.10	0.36	0.84	0.80	0.80	122

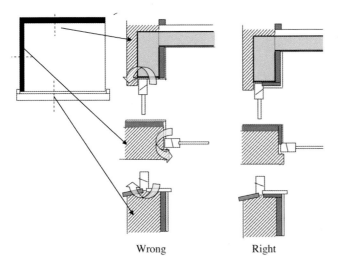

Wrong   Right

Figure 9.17. Inside insulation, solving the window reveals.

The value found surpasses the clear wall one by 84 to 122%. Or, inside insulation does not allow excellent performance due to a disappointing efficiency (ranges from 0.45 to 0.54). For the concrete structure, a whole wall thermal transmittance 0.6 W/(m²·K) requires a 24 cm thick insulation, albeit 5 cm should suffice if the efficiency was 1.

Also other details demand appropriate attention. Outer door and window reveals for example must be insulated up to the window- and doorframe (Figure 9.17). If not, another marked thermal bridge is created.

Massive wall airtight, indoor air washing

Indoor air washing behind the insulation may further increase the whole wall thermal transmittance. In fact, enthalpy losses add to conduction from indoors. Together they define the heat flow across the massive wall. With the air layer simplified to a thermal resistance and the ordinate $z = 0$ coinciding with the ceiling the balance becomes (see Figure 9.18):

$$\frac{\theta_i - \theta_c}{R_i^c} + \frac{\theta_e - \theta_c}{R_c^e} = c_a \, G_a \, \frac{d\theta_c}{dz} \tag{9.5}$$

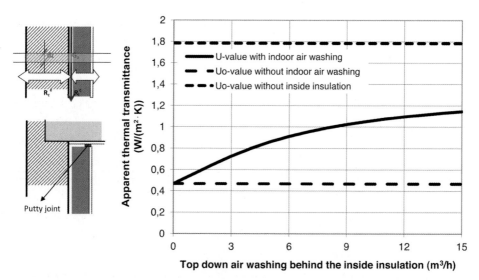

Figure 9.18. One-brick masonry wall insulated inside. Upside down indoor air washing behind the insulation, apparent thermal transmittance as a function of the washing flow (monthly mean inside temperature 21 °C, monthly mean outside temperature 3.2 °C).

where R_i^c is the thermal resistance between inside and air layer, R_c^e the thermal resistance between air layer and outside, θ_c the temperature in the air layer, c_a the specific heat of air and G_a the washing flow in kg/s. For $z = 0$, the temperature $\theta_{c,o}$ in the air layer hardly differs from the one inside θ_i (z-axis pointing top-down). Temperature in the air layer then becomes:

$$\theta_c = \theta_{i\infty} + (\theta_i - \theta_{i\infty}) \exp\left(-\frac{R_i^c + R_c^e}{c_a\, G_a\, R_i^c\, R_c^e} z\right) \tag{9.6}$$

The driving force behind is buoyancy:

$$\Delta p_T = 3450 \left[\int_0^h \frac{dz}{T_c} - \frac{h}{T_i} \right] \tag{9.7}$$

The washing flow itself follows from the equilibrium between driving force, friction and local losses in the air layer:

$$\Delta p_T = C_1\, G_a + C_2\, G_a^2 \tag{9.8}$$

with $C_1\, G_a$ the friction for laminar flow and $C_2\, G_a^2$ local dynamic loss. The constants C_1 and C_2 are case-related. Solving (9.6)–(9.8) requires iteration. In fact, the buoyancy force as well as the washing flow change with temperature in the air layer, whereas temperature depends on the washing flow. In the absence of any thermal bridge effect, the apparent thermal transmittance looks like:

$$U_{o,\text{eff}} = \frac{1}{h(\theta_i - \theta_e)} \int_0^h \frac{\theta_c - \theta_e}{R_c^e} dz = U_o \left[1 + \frac{R_i^c}{R_c^e} F(a, h)\right]$$

with:

9.3 Massive walls with inside insulation

$$F(a, h) = \left[\frac{1-\exp(-a\,h)}{a\,h}\right] \text{ en } a = \frac{R_i^c + R_c^e}{R_i^c\, R_c^e\, c_a\, G_a} \tag{9.9}$$

Figure 9.18 gives the result for one-brick masonry, insulated at the inside with 6 cm mineral wool. Without air washing, clear wall thermal transmittance reaches 0.47 W/(m² · K). With air washing, the worst case considered, assuming 15 mm wide joints at skirting and ceiling level and an air layer of 15 mm behind the insulation, gives 1.15 W/(m² · K), an increase of 145%! In practice, 15 mm is not exceptional. The apparent thermal transmittance also does not increase linearly with the washing flux. Since buoyancy increases as the air layer turns colder, air washing will have the largest impact in winter. Also the more severe the thermal transmittance requirements, the more detrimental washing is. Happily, avoidance is easy; a putty joint between inside lining and floor or ceiling suffices. Besides indoor air washing, even other airflow patterns may develop. The stud-based type for example is quite sensitive to air looping in and around the mineral wool insulation.

Transient response

Table 9.7 lists dynamic thermal resistance, temperature damping and admittance of one-brick masonry with or without a 4 or 16 cm EPS/12.5 mm gypsum board sandwich as inside insulation. Although the dynamic thermal resistance increases substantially, temperature damping with 4 cm EPS hardly does. Even 16 cm EPS fails to push the value beyond 15. Worse is that, independent of insulation thickness, admittance reflects lightweight outer walls. Or, inside insulation disconnects the massive wall as a heat storage medium from indoors. Requirements such as temperature damping beyond 15 and admittance larger than half the surface film coefficient indoors disqualify the solution.

Table 9.7. Inside insulation: dynamic thermal resistance, temperature damping and admittance of a massive wall without and with inside insulation (1-day period).

Outer wall Inside EPS thickness d+ gypsum board lining	Temperature damping		Dynamic thermal resistance		Admittance	
	m² · K/W	Phase, h	m² · K/W	phase, h	W/(m² · K)	phase, h
One brick thick masonry	3.6	7.4	0.85	6.3	4.2	1.1
Idem, d = 4 cm	4.1	9.4	5.1	7.1	0.80	2.3
Idem, d = 16 cm	10.5	12.6	16.3	8.3	0.65	4.3

A poor outer wall transient performance nevertheless should not be a problem as long as the partition walls and floors can store heat and the glazed surface is moderate. Even with more glass, solar shading and night ventilation may keep summer comfort acceptable in moderate climates. Only when the inside partitions also get an insulation layer, as is the case for concrete cast in EPS formwork, does overheating become more likely.

Moisture tolerance

The discussion that follows considers wind driven rain and interstitial condensation only. Rising damp is easily excluded by inserting a damp proof course just above grade in outer and partition walls and for the outer walls above protrusions were run-off collects. Drying of building moisture is problematic only when the outer wall has a vapour retarding outside finish.

The response then resembles the one of a wall wet by rain. For mould and surface condensation reference is made to thermal bridging. One remark: as inside insulation allows frost all over the massive wall, water supply lines, heating pipes and guiding rods for electrical wires should be kept out. In existing buildings, all must be removed and channelled in partition walls or mounted in a utility cavity between insulation and inside lining.

Wind driven rain

With inside insulation the wall's outside surface operates as a drainage plane and the wall as a buffering volume. Buffering prevails until the outside surface becomes capillary wet (w_c). Then run-off starts. For masonry with high capillary water absorption coefficient (A) and high capillary moisture content (w_c), only long lasting rain causes that reversal. Anyhow, before it gets that far, the rain front progresses in the masonry, the depth (x) reached being approximated as:

$$x = \frac{D_w \left(w_{cr} - w_H \right)}{g_{ws}} \left[\sqrt{1 + \frac{2 \, g_{ws}^2 \, t}{D_w \left(w_{cr} - w_H \right)^2}} - 1 \right] \tag{9.10}$$

where w_{cr} is critical moisture content, g_{ws} the mean wind driven rain intensity and D_w moisture diffusivity of the masonry, given by $D_w = A^2 / (w_c^2 - w_{cr}^2)$. As long as buffering goes on, moisture content at the outside surface increases:

$$w = w_{cr} + g_{ws} \, x / D_w \tag{9.11}$$

Once capillary wet and run-off started, buffering goes on until the massive walls inside surface reaches capillary moisture content. The time in which run-off forms compared to the moment precipitation began is called the turnover moment. With run-off, seepage risk increases, ending with rainwater running off in the air layer between massive wall and insulation. Shrinkage cracks across the masonry, badly sealed joints and protrusions where run-off collects and forms water heads are the vulnerable places. That is why sills, copingstones and reliefs must slope away from the wall. For through cracks, the time needed for water to penetrate is $D^2/(2.35 \, d)$ with D wall thickness and d crack width, both in m. Crossing a 0.1 mm wide through crack in a 14 cm thick wall for example only requires 83 seconds! Buffering during run-off is approximated as follows:

Rainwater did not reach the inside surface of the massive wall yet ($x < D$):

$$t' = \left[\frac{x \left(w_c + w_{cr} \right)}{2 A} \right]^2 \qquad x = \frac{2 A}{w_c + w_{cr}} \sqrt{t - t_f + t'}$$

$$m_T = \frac{w_c + w_{cr}}{2 \, x} + \frac{w_{H,95\%} - w_{H,50\%}}{2 \, (D - x)} \tag{9.12}$$

Rainwater reached the inside surface of the massive wall ($x = D$):

$$w^{t+\Delta t} = w_{cr} + \frac{D_w \left(w_c - w^t \right) x}{d^2} \Delta t$$

$$m_T = \frac{w_c + w^{t+\Delta t}}{2 D} \tag{9.13}$$

9.3 Massive walls with inside insulation

In these equations, t_f is the turn-over moment, t' the corrected turn over moment and t total time, the three relative to the time precipitation started, and x is wet thickness. In case the rainwater reached the inside surface of the massive wall before the outside is capillary wet, buffering is approximated using the equations:

$$\Delta w = \frac{D\, g_{ws}}{D_w} \Delta t$$

$$w_{x=0}^{t+\Delta t} = w_{x=0}^{t} + \Delta w \qquad w_{x=D}^{t+\Delta t} = w_{x=D}^{t} + \Delta w \qquad (9.14)$$

$$m_T = \frac{w_{x=0}^{t+\Delta t} + w_{x=D}^{t+\Delta t}}{2D}$$

Figure 9.19 gives results for a 20 cm thick concrete and a 20 cm thick brick wall. Differences are striking. Concrete humidifies slowly. Even after 4 hours of intensive wind driven rain, the water front only advanced 2 cm, while in the brick wall it already touched the inside surface.

Looking to run-off, the same difference is apparent. Even after 4 hours of moderate wind driven rain, no water film developed on the brick wall, whereas on the concrete wall, fingering run-off started after a couple of minutes, see Figure 9.20.

As soon as it stops raining, drying starts, though in winter slower with than without inside insulation. If the wall is not capillary moist across its thickness, the wet zone simultaneously expands in the direction of the insulation. At the same time, low diffusion thickness of that insulation package allows water vapour produced indoors to deposit against the massive wall's inside surface. As Figure 9.21 shows, for a 40 cm thick massive wall, both mechanisms elevate the wall's wetness compared to no inside insulation, increasing frost damage risk, the likelihood of salt efflorescence and sensitivity for algae growth.

These modelling results were verified experimentally by testing five 60 × 60 cm large one-brick wall samples, plastered inside. One sample remained non-insulated, the other got inside insulation, with 5 cm mineral wool ($\mu = 1.2$), 5 cm PUR ($\mu - 21$), 5 cm XPS ($\mu = 122$) and 5 cm mineral wool plus vapour retarder ($\mu d = 20$ m) respectively. The insulated ones were finished with a gypsum board inside lining. First the five were capillary saturated, after which they were subjected in a hot box during 110 days to a winter outdoor climate ($\theta_e = 0.4$ °C, $p_e = 520$ Pa) and a quite humid indoor climate ($\theta_i = 23.8$ °C, $p_i = 2125$ Pa).

Figure 9.22 gives the drying curves. First all five dry. Quite soon however, the non-insulated wall sees its moisture content increasing again due to surface condensation. The same happens with the wall insulated with mineral wool, the reason now being interstitial condensation in the interface with the plastered masonry. Only the walls insulated with PUR and XPS and the one with a vapour retarder go on drying. After 110 days, moisture content was measured. In all cases, the plaster was still capillary (PUR, XPS, vapour retarder) or above capillary wet (mineral wool). There, also the average moisture content in the masonry was high, 143 kg/m³, i.e. beyond critical (≈ 100 kg/m³).

Even more happens in moderate and cold climates. At the start of the warm season, sunny days lift water vapour saturation pressure in the wet masonry to values far above the vapour pressure indoors. That allows solar driven vapour to cause condensation at the insulation-side of the vapour retarder or inside lining and in the insulation package, see Figure 9.23. The risk this happens increases the more vapour permeable the insulation and the more vapour retarding the inside lining is. In fact, the amount of condensate deposited is inversely proportional to the diffusion thickness of the insulation.

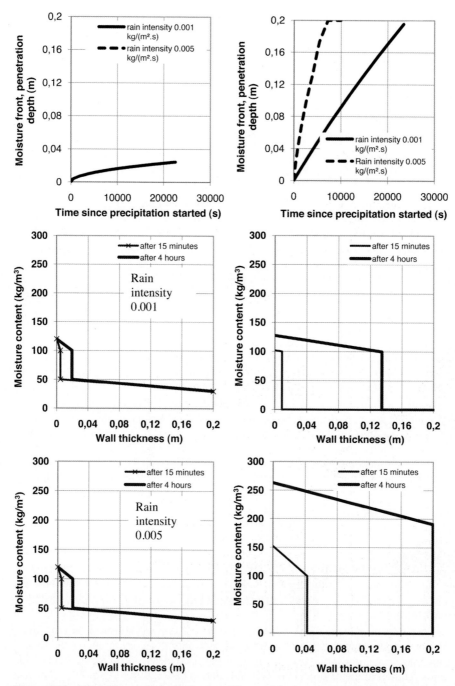

Figure 9.19. Wind driven rain, buffering by a 20 cm thick concrete and 20 cm thick brick wall, both insulated at the inside. Penetration depth and moisture profile calculated with the simple model just explained.

9.3 Massive walls with inside insulation

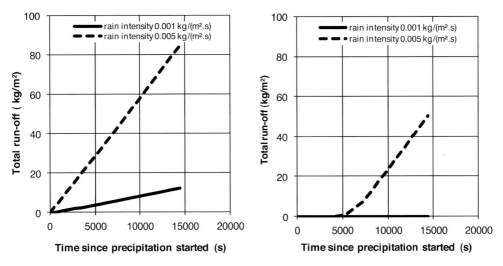

Figure 9.20. Wind driven rain impinging on a 20 cm thick cast concrete and a 20 cm thick brick wall, both insulated at the inside. Run-off.

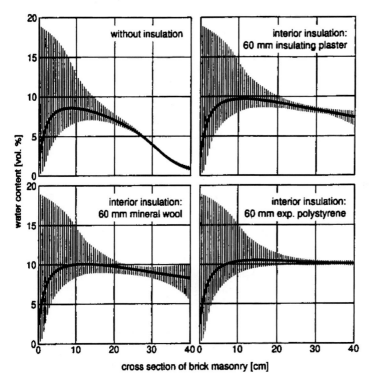

Figure 9.21. Moisture in 40 cm thick brick wall without and with inside insulation (60 mm insulating plaster, 60 mm mineral wool, 60 mm EPS). The full line is the annual mean moisture content, the shaded surface the annual fluctuation (calculated with WUFI$^{(C)}$ for Holzkirchen, Bavaria, Germany).

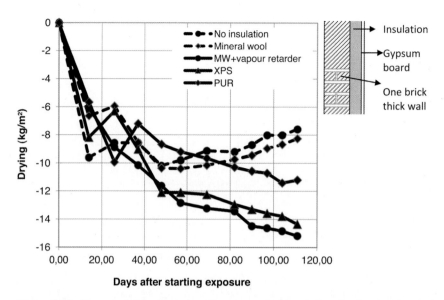

Figure 9.22. Five one-brick masonry walls, one non-insulated, the other insulated at the inside. Drying after capillary wetting. Winter conditions in the cold box ($\theta_e = 0.4$ °C, $p_e = 520$ Pa), humid indoor climate in the hot box ($\theta_i = 23.8$ °C, $p_i = 2125$ Pa).

Table 9.8. Inside lining diffusion thickness pivot for excluding solar driven condensation.

Insulation material	Maximal diffusion thickness (m)		
	Indoor climate class 1	Indoor climate class 2	Indoor climate class 3
Mineral wool	←	Condensation in MW	→
EPS	0.50	0.48	0.44
PUR	0.70	0.66	0.61
XPS	2.90	2.70	2.49
Cellular glass	←	No requirements	→

Table 9.8 summarises the inside lining requirements that exclude solar driven condensation in a moderate climate for the indoor climate classes 1, 2 and 3 of EN ISO 13788. Only XPS and cellular glass allow using a more vapour retarding inside finish. All other insulation materials demand a vapour permeable one. Indoor climate class 4 and 5 are not included. As will be shown below, inside insulation is not a solution there.

Mineral wool gives such high condensation deposit that rain buffering is not permitted, if applied against a one-brick outer wall. In other words, the wall must be rain-protected with stucco or a siding. Some have proposed a water repellent treatment as alternative. Such treatment however may fail, making things worse as treated brickwork dries slower!

9.3 Massive walls with inside insulation

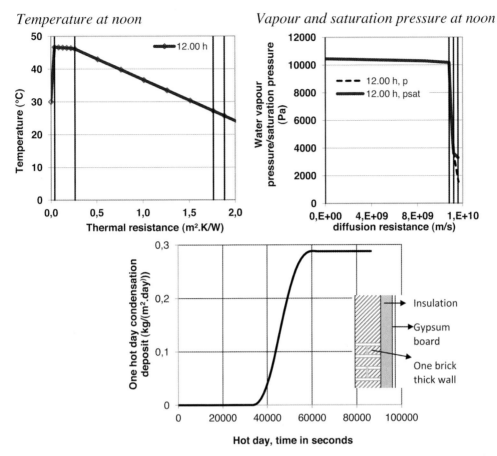

Figure 9.23. One-brick masonry insulated inside with 5 cm mineral wool finished with gypsum board lining. Interstitial condensation in the wool and at the backside of the lining during a hot summer day.

Interstitial condensation

In general, inside insulation conflicts with the rule for airtight assemblies that the best insulating and most vapour permeable layer should be located outside and the worst insulating and least vapour permeable layer inside. The question then becomes, to what extent is interstitial condensation acceptable? An answer has been sought experimentally by testing five one-brick 60 × 60 cm large walls, one not insulated and the other four insulated at the inside with 5 cm mineral wool ($\mu = 1.2$), 5 cm PUR ($\mu = 21$), 5 cm XPS ($\mu = 122$) and 5 cm mineral wool plus vapour retarder ($\mu d = 20$ m), the four finished with a gypsum board internal lining. During 110 days the five walls were subjected to a winter outdoor climate ($\theta_e = 0.4$ °C, $p_e = 520$ Pa) and a humid indoor climate ($\theta_i = 23.8$ °C, $p_i = 2125$ Pa) in a hot box. Figure 9.24 shows moisture uptake.

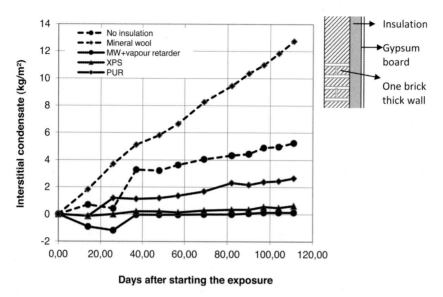

Figure 9.24. Five one-brick walls, one not insulated, the other four insulated at the inside. Interstitial condensation under winter outdoor conditions ($\theta_e = 0.4$ °C, $p_e = 520$ Pa) and humid indoor conditions ($\theta_i = 23.8$ °C, $p_i = 2125$ Pa).

That looks inversely proportional to the vapour thickness of the insulation layer. Mineral wool gives a deposit of 117 g/(m² · dag), PUR one of 29 g/(m² · dag), XPS 7 g/(m² · dag) and mineral wool with vapour retarder 1 g/(m² · dag).

Air washing makes things even worse. This again was evaluated experimentally. In a first step a one-brick masonry wall, with the inside surface plastered, got an airtight inside insulation package, 5 cm mineral wool, a vapour retarder with diffusion thickness 4 m and a gypsum board inside lining. After 41 days of exposure to 0 °C, 550 Pa and 16 °C, 1360 Pa, the original plaster reached a moisture content corresponding to 70% relative humidity. In a second step, indoor air washing was introduced by remounting the insulation, now with a 5 mm air layer behind it and open joints at top and bottom. After 60 days of exposure to the same conditions, moisture content in the original plaster reached values between 220 and 250 kg/m³, while showing mould over its full height. Of course, one could claim indoor air washing is a construction defect, so these observations are too negative. Practice however shows the washing risk is real.

<u>Wall with rain repelling finish</u>

The question anyhow remains how acceptable interstitial condensation is in case the wall/inside insulation combination is airtight, free of indoor air washing and rain protected. The answer differs depending on indoor climate class, capillary properties of the massive wall, diffusion resistance of the exterior finish, frost resistivity of the massive wall and yes or no yearly accumulation of condensate. Besides, for stud-based types, the timber should be treated for usage outdoors. Otherwise, the highest monthly mean relative humidity at the timber must stay below 90%.

9.3 Massive walls with inside insulation

No yearly accumulating condensate

If for one or the other reason the massive wall shows no capillary suction, then the annual deposit should be limited to a maximum of 100 g/m². If it does show, then the condensation flow rate m_c/t with t the length of the condensation period and m_c total winter deposit is sucked by the wall, causing a wet thickness given by Equation (9.10) but now expressed by:

$$x = \frac{D_w \, w_{cr} \, t}{m_c} \left[\sqrt{1 + \frac{2 m_c^2}{D_w \, w_{cr}^2 \, t}} - 1 \right]$$

Moisture content at the inside surface becomes:

$$w = w_{cr} + \frac{m_c}{t} \frac{x}{D_w}$$

If that value exceeds capillary moisture content, which is quite unlikely, water droplets will form. Once total droplet weight passes 100 g per m², they coagulate and run-off starts. In case surface moisture content stays below capillary, no problems should be expected.

With a capillary active insulation layer, that shows higher suction potential than the massive wall's surface, the insulation and not the wall turns moist in winter with a net increase in clear wall thermal transmittance as a consequence.

Yearly accumulating condensate

Not allowed with stud-based inside insulation. If for the sandwich type the massive wall's inside surface is not capillary, yearly accumulation is out of question. When instead the massive wall and its inside surface are capillary, a limit state analysis at yearly mean inside and outside conditions should be done. If at equilibrium the moisture front is located in a non-capillary outside finish, then the wall will end capillary wet with run-off, which is unacceptable. If instead the moisture front stabilizes in the massive wall, then the moisture content at its inside surface will near:

$$w = w_{cr} + \frac{g_c \, x}{D_w} \tag{9.15}$$

with g_c condensing flow rate and x thickness of the moist layer against the massive wall's inside surface. Should moisture content at the massive walls inside surface exceed capillary, which is quite unlikely but not impossible, then in reality humidification will stop at capillary with further deposit running off, again not acceptable. Also capillary moisture content in masonry lacking frost resistance cannot be tolerated. The only acceptable situation thus is moisture content below capillary across the wet zone in frost resistant masonry. Table 9.9 summarizes the discussion in terms of vapour retarder quality needed in a moderate climate.

Table 9.9. Inside insulation, decision array in terms of vapour retarder needed (moderate outside climate, wall rain protected, masonry frost resisting).

Indoor climate class	Isolation	Vapour retarder (E1: 2.5 m < $(\mu d)_{eq} \leq 5$ m, E2: 5 m < $(\mu d)_{eq} \leq 25$ m)? Inside insulation type				
		All	Sandwich		Stud based	
			Massive wall			
		Non capillary	Capillary		Capillary	
			Concrete	Other	Concrete	Other
1	MW	E1[2]	None	None	> 0.1 m[1]	None
	EPS, PUR	None	None	None	None	None
	XPS	None	None	None	None	None
2	MW	E1	E2[3]	None	E2	E1
	EPS, PUR	E1	E1	None	E1	None
	XPS	None	None	None	None	None
3	MW	E2	E2	E1	E2	E2
	EPS, PUR	E2	E2	No	E2	None
	XPS	None	None	None	None	None
4, 5		**Never apply inside insulation**				

[1] μd inside finish
[2] $(\mu d)_{eq} \leq 5$ m
[3] 5 m < $(\mu d)_{eq} \leq 25$ m

Wall without rain repelling finish

For a rather thin wall (one brick) looking to the main wind-driven rain direction, the situation becomes quite complex. Of course, a limit state analysis allows guessing at which thickness the wall will stay wet at the insulation side. Certainly solar driven vapour flow makes a vapour retarding layer between vapour permeable insulation and inside finish, as typically seen in stud-based types, a bad solution.

Conclusion

Only two 'safe' options are left:

1. Rain-protecting the wall before adding any inside insulation type, included a correct vapour retarder when needed
2. Using sandwich types with an insulation material with high enough vapour resistance factor, such as XPS, to exclude problems with solar driven vapour flow

In both cases, indoor air washing must be excluded by correctly sealing the inside finish, while rain-protecting presumes stuccoing, siding or, less safe, treating the masonry's outside surface with water repellent siloxanes. Only do the last when the mortar joints are correctly pointed and do not show micro cracking.

9.3 Massive walls with inside insulation

As mentioned, a third solution advanced to date is capillary insulation materials. Such insulation however may turn wet by interstitial condensation and by rain absorption if applied on walls without rain repelling outside finish and no capillary break behind the insulation. In that case, solar driven vapour flow may still moisten the insulation if finished inside with a vapour retarding lining. Laboratory controls also showed that many insulation boards that are claimed to be capillary active, hardly are.

Thermal bridges

Table 9.8 gives the lowest temperature ratio for a concrete floor/concrete outer wall insulated inside junction and a concrete partition/same outer wall junction. The calculations assumed a cupboard standing against the outer wall. In both cases, the values found stay far below 0.7, pushing mould risk far above 5%.

Table 9.8. Concrete construction, inside insulation (6 cm MW), lowest temperature ratio.

Thermal bridge	Construction	f_{hi}
Unavoidable thermal bridge		
Outer wall/inside wall		0.50
Floor support		0.55
Avoidable thermal bridge		
Window reveal	Non insulated	0.38
	Insulated	0.50

All details demand extra care. If for example one forgets insulating window and door reveals in a concrete outer wall, very low temperature ratios and high mould risk can be expected. Insulating the reveals increases the temperature ratio at the window but, more importantly, limits the zone with low values (Figure 9.25).

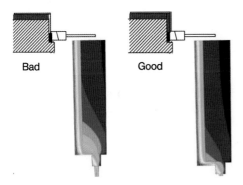

Figure 9.25. Concrete wall insulated inside, window reveal, isotherms.

9.3.2.3 Building physics: acoustics

Inside insulation makes a composite from a massive wall with the inside lining as resilient layer and the thermal insulation as elastic support. In case that support is stiff, as most synthetic foams are, sound transmission loss drops compared to the non-insulated wall – see Figure 9.26. As for envelopes, glazing anyhow remains the acoustically weak link, so that drop is not alarming.

More annoying is that in terraced houses outer walls insulated inside behave as flanking transmission paths for the party walls. As a rule of thumb inside insulation with EPS lowers the sound transmission loss between adjacent houses by up to 10 dB. In light of the 52 dB or higher requirement, this argues against inside insulation.

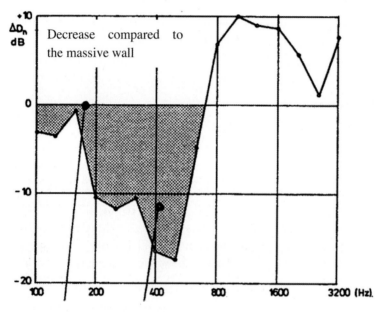

Figure 9.26. Sound transmission loss: decrease by insulating a massive wall inside with a EPS/gypsum board sandwich.

9.3.2.4 Durability

Once initial shrinkage is over, temperature and humidity are left as causes of differential movement. Table 9.10 gives the maximum and minimum temperatures to be expected in a moderate climate at the outside and inside surface of a south-west looking one-brick outer wall without and with inside insulation.

With inside insulation, the wall turns colder in winter. In summer, temperatures only increase a little. For a whole year, temperature differences increase, whereas in winter frost moves into the insulation to stay there for days during cold spells, even when the outer part of the wall defrosts each afternoon at the sunny sides.

9.3 Massive walls with inside insulation

Table 9.10. One-brick wall without and with inside insulation, temperatures (moderate climate, cold winter and hot summer day, SW looking, se = outside surface, 1= interface wall/insulation).

Wall		Temperatures Cold winter day		Temperatures Hot summer day	
		Minimum	Maximum	Minimum	Maximum
One-brick wall without inside insulation	se	−11.4	7.4	20.9	44.9
	1	8.2	10.8	22.9	31.3
Idem, 4 cm MW, gypsum board lining	se	−13.5	6.0	21.8	45.3
	1	−9.1	−1.1	24.7	36.2
Idem, 16 cm MW, gypsum board lining	se	−14.2	7.1	21.3	47.0
	1	−10.5	−1.9	24.8	36.9

Load bearing outer walls so turn periodically warmer and colder than the inside partitions. Warmth induces compression in the outer and tension plus shear in the adjacent partitions walls. Cold causes tension in the outer and compression in the adjacent partitions walls. Inside insulation enlarges these changes with a higher risk tensile stress will exceed ultimate strength in both, causing 45° cracks in the partitions and horizontal and vertical cracks in the outer walls. In medium and high rises, that cracking can certainly cause problems. There is one redeeming factor. Thanks to the insulation inside, temperature differences across the outer walls diminish and bending moments in the contacts with decks and inner partitions drop. Increased temperature fluctuations by inside insulation also enlarge thermal movement of concrete lintels with cracking between lintel support and masonry as the most probable result.

In case the outer wall becomes wet, hygric movement increases. Assume the wall is freely supported and rain saturates the masonry capillary over a thickness of d_w. Mean elongation of the wall then becomes (ε_c: hygric swelling between dry and capillary moist):

$$\varepsilon = \varepsilon_c \frac{d_w}{d} \tag{9.16}$$

When the centre of the capillary wet zone is not midway in the wall ($d_w \neq d/2$) bending complements that elongation:

$$\varepsilon_b = \frac{3\,\varepsilon_c\,d_w\,(d-d_w)\,y}{d^3} \tag{9.17}$$

If both are hindered, the dry zone develops tension and the wet zone undergoes compression.

9.3.2.5 Fire safety

Synthetic foams must be finished with a fire retarding non-burnable inside lining. Gypsum board is a good choice, although the boards turn warmer during fire with than they do without inside insulation, reducing retarding that way. The whole insulation package should of course be mounted as airtight as possible. If not, thermal stack may suck hot smoke between wall and insulation, causing the foam to melt prematurely.

9.3.2.6 Global conclusion

Inside insulation appears to be synonymous to 'annoying consequences':

- Unavoidable thermal bridges put a limit to what clear wall thermal transmittance is a fair objective. That seems a value 0.6 W/(m² · K). For low energy buildings inside insulation is no option
- Overlooking stack induced indoor air washing further decreases insulation efficiency
- Inside insulation deprives massive outer walls from their thermal capacity, imposing stricter requirements in terms of passive measures to guarantee good summer comfort
- Inside insulation complicates moisture tolerance, with rain absorption and not interstitial condensation as the big bugaboo. That in winter a capillary active outer wall insulated inside stays wetter and colder than a non-insulated one increases frost damage risk, fosters salt efflorescence, sometimes salt degradation, and favours algae growth
- Unavoidable thermal bridges increase mould risk
- Using stiff insulation degrades sound insulation
- The higher hygrothermal load between winter and summer induces more stress and strain in the massive outer and adjacent partition walls. The fabric may show worse cracking than without inside insulation

9.3.3 Design and execution

Although inside insulation is a problematic technique, sometimes an alternative is lacking, for example when the clear wall thermal transmittances are really too high and the building's look may not change as is the case for landmark buildings. Before deciding on applicability, a risk analysis must be performed. How large the probability is all drawbacks will affect enclosure durability in the end differs from case to case. Thicker masonry for example is less sensitive to overall wetting than a thinner one. Walls that are barely capillary also pose less risk. Outer walls may be protected against wind-driven rain by the surroundings, etc.

Two quite safe choices are:

- Inside insulation combined with rain repellent outside finish (stucco, outside siding, sometimes a water repellent treatment)
- Sandwich systems composed of insulation boards with high vapour resistance factor and the massive wall built of frost resisting low-salt blocks and mortar

Correct mounting of the insulation is critical. The massive wall must be free or, in case of a retrofit, freed of water pipes, hydronic central heating pipes and guiding rods for electrical wires. The first two freeze in winter whereas the third may figure as an air leak between the inside and the cold massive wall with condensation in the rods as a consequence. When for the one or the other reason guiding rods for electrical wires are needed in the enclosure, the best way is to leave a ducting cavity between the insulation and the inside lining and mount them there (Figure 9.27).

9.3 Massive walls with inside insulation

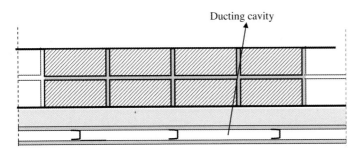

Figure 9.27. Inside insulation, ducting cavity.

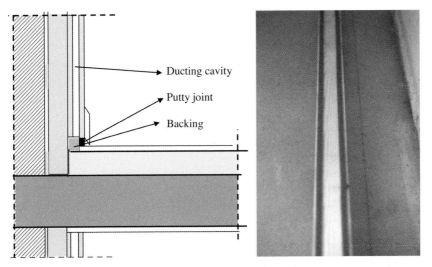

Figure 9.28. Inside insulation: closing the floor joint. Left: by casting the screed against the insulation and sealing the joint between gypsum board and floor with a mastic compound, right: by adhering swelling strips to the bottom plate.

With sandwiches, after attaching the storey-high elements, joints at floor and ceiling must be carefully sealed. An alternative for the floor joint consists of first mounting the elements after stripping the gypsum board over a height somewhat more than screed thickness and then casting the screed. Stripping avoids cleaning water from being sucked by the gypsum board. In case of a timber stud solution, swelling strips between bottom plate and floor and top plate and ceiling (Figure 9.28) best replace sealing. Finishing the gypsum boards is done according to the manufacturer's instructions (covering the joints with glass fabric, equalizing with filler, etc).

Window and outer door reveals, lintels and sills, all have to be insulated. That the insulation thickness is less there than elsewhere, is not a problem. The joints between insulation and window or door assemblies must also be sealed.

9.4 Massive walls with outside insulation

9.4.1 In general

Also with outside insulation, stud solutions are applied. Timber or steel studs are screwed against the wall, the bays in between filled with mineral wool boards, the whole covered with a wind retarding foil, after which batten and laths are fixed and the wall finished with timber siding, slates, tiles, panels, etc. Alternatively, stainless steel L anchors with vertical timber beams screwed against them (Figure 9.29) replace the studs.

Stuccoed outside insulation, called EIFS, is an alternative. With it, the insulation boards (mineral wool, EPS, PUR, cellular glass) are glued against the wall and, when needed, additionally fixed with synthetic nails. The insulation then gets a base render with a glass fabric embedded. Beforehand one equalizes the insulation boards and reinforces all corners with stainless steel stucco guides. Decorative stucco, such as mineral, synthetic or silicone, finishes the whole – see Figure 9.30. Some systems have self-bearing stucco. Sometimes the insulation boards are attached using a system of sunken stainless steel profiles.

Figure 9.29. Massive walls with outside insulation: stud systems.

Figure 9.30. Massive walls with EIFS
(on the left: while insulating, in the middle: once finished, on the right: a view of the assembly).

9.4 Massive walls with outside insulation

9.4.2 Performance evaluation

9.4.2.1 Structural integrity

Outside insulation systems are not load bearing. Anyhow, with EIFS, the insulation has to transmit the stucco weight to the massive wall, while its bond strength must be high enough to withstand wind suction. Suction in fact redistributes between stucco and wall proportional to either's air resistance. In that process, due to the joints between boards or their air permeabililty, the insulation hardly intervenes. If stucco and wall have equal air resistance, then both take half the wind suction with the insulation transmitting the stucco half. If the massive wall is more airtight, then it takes it nearly all, which is the safest option.

9.4.2.2 Building physics: heat, air, moisture

Air tightness

At first sight, outside insulation does not pose problems. Awareness however cannot hurt. When the massive wall consists of fair face concrete blocks and a stud based outside system with mineral wool is applied, air-tightness is not guaranteed. Mineral wool in fact is air permeable, just like timber sidings, slates, tiles and panels, see Table 9.11.

But there is more. When EIFS is not sealed at top and bottom and the insulation boards are not glue-bonded at their perimeter, then the airflow pattern of Figure 9.31 become likely. In stud based systems with vented or ventilated cavity behind the cladding and carelessly mounted insulation boards, three flow patterns may develop: outside ↔ cavity, outside ↔ behind the insulation and cavity ↔ behind the isolation. Covering the insulation with a wind-tight foil stops the second and third pattern. We assume that in what follows.

Table 9.11. Air permeance: one-stone wall of concrete blocks and a few outside finishes.

Wall, outside finish	Execution	Air permeance $K_a = a \Delta P_a^{b-1}$, kg/(m²·s·Pa)	
		$a (\cdot 10^{-4})$	$b - 1$
One-stone wall	Fair face concrete blocks	1.71	−0.13
Slates	Overlapping	14.0	−0.3
Timber siding	Groove and tongue joints	4.1	−0.32

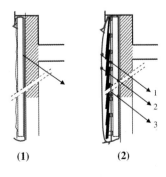

Figure 9.31. Outside insulation:
(1) airflow behind the insulation with EIFS-systems,
(2) the three possible air flow patterns (1, 2, 3) in stud-based systems with cavity between insulation and cladding (massive wall airtight).

Thermal transmittance

Four variables fix the clear wall thermal transmittance of outside insulation systems: type and thickness of the massive wall and type and thickness of the insulation. Because too thick outer walls diminish the net area for a same out to out area, which means less usability for a given investment, or requires larger out to out area for the same net area, which means the same usability but higher investment, massive wall thickness is best limited to 19 cm. A value 0.1 to 0.6 W/(m² · K) with EIFS then requires the insulation thicknesses of Table 9.12.

Table 9.12. Clear wall thermal transmittance: insulation thicknesses when applying EIFS.

Massive wall	Uo-value W/(m² · K)	Insulation thickness m			
		MW	PUR	EPS	CG
Perforated fast bricks, *d* = 14 cm	0.6	4	3	5	5
	0.2	17	13	18	20
	0.1	35	27	38	42
Light weight fast bricks, *d* = 14 cm	0.6	4	3	4	4
	0.2	16	12	17	19
	0.1	35	26	37	41
Perforated fast bricks, *d* = 19 cm	0.6	4	3	4	5
	0.2	16	12	18	19
	0.1	35	26	38	41
Light weight fast bricks, *d* = 19 cm	0.6	3	2	3	4
	0.2	15	12	17	18
	0.1	34	26	37	40

Types and thicknesses of the massive wall hardly have an impact, whereas the effect of the insulation thickness on the clear wall thermal transmittance is overwhelming. Up to a value of 0.2 W/(m² · K) total thickness remains acceptable: including inside plaster and outside stucco 28–33 cm for PUR and 35–40 cm for cellular glass. Except for PUR, the step to 0.1 W/(m² · K) boosts thicknesses to uneconomical levels: 50 cm with mineral wool and 56 cm with cellular glass.

Window bays, outer door bays and balconies may induce thermal bridging. Figure 9.32 illustrates this for a 44 cm high lintel/concrete floor slab combination, a window reveal and sill, a floor support, a balcony and a roof edge in case of a 19 cm thick masonry wall, a 10 cm thick insulation layer, a timber inner sill, a fibre cement outer sill and the window parallel to the outside surface of the masonry.

As Table 9.13 underlines, geometrical effects lift the linear thermal transmittances for two of the three window-related details to high values. The isotherms clarify why. By using the insulation as rebate, steep temperature gradients develop at the reveal. The balcony also performs badly as a rigid coupling to the floor is needed to withstand the support moment, which is why concrete slabs often cantilever to form balconies.

9.4 Massive walls with outside insulation

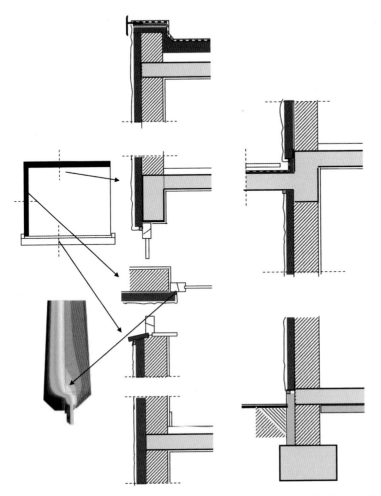

Figure 9.32. Outside insulation, details. Minimizing thermal bridging at balconies can be difficult. Special thermal cut systems that guarantee continuity of the bending and shear steel bars create possibilities.

Table 9.13. Details shown in Figure 9.31, linear thermal transmittance (ψ).

Detail	ψ W/(m·K)
Lintel (44 cm high)	0.57
Reveal	0.38
Outer sill (wooden inner sill)	0.19
Floor support	0.006
Balcony (without thermal cut between it and the floor slab)	0.84
Roof edge	0.11

Transient response

Table 9.14 recaptures the one-brick masonry wall of Table 9.7, insulated at the inside, now giving temperature damping, dynamic thermal resistance and admittance if EIFS-insulated with 4 and 16 cm thick EPS boards. Also, the walls of Table 9.12 are included, though only with mineral wool as insulation material. 4 cm EPS outside gives eight times more temperature damping and a five times higher admittance than 4 cm EPS inside. With 16 cm EPS outside, temperature damping even increases by a factor of twelve. In addition, the dynamic thermal resistance gains, though only by a factor of two.

Reaching a temperature damping value of 15 constitutes no problem. Only lightweight fast brick walls with a thin insulation layer perform worse. Admittances of 3.9 W/(m² · K) are harder to exceed. The one-brick wall succeeds, fast bricks come close but lightweight fast bricks fail. The admittance in fact mainly depends on the contact coefficient at the inside surface. With outside insulation, the best is opting for heavy walls in cast concrete or sand-lime stone. That conclusion however must be put in perspective. As mentioned before, glass type, glass orientation and area, with or without solar shading, ventilation strategy and the admittance of the partition walls, floor and ceiling are far more important than the properties of the opaque facade.

Table 9.14. EIFS: temperature damping, dynamic thermal resistance, admittance (1-day period).

Wall	Temperature damping		Dynamic thermal resistance		Admittance	
	–	Phase, h	m² · K/W	Phase, h	W/(m² · K)	Phase, h
One-brick masonry	3.6	7.4	0.85	6.3	4.2	1.1
4 cm EPS inside, gypsum board	4.1	9.4	5.1	7.1	0.8	2.3
16 cm EPS inside, gypsum board	10.5	12.6	16.3	8.3	0.65	4.3
EIFS, 4 cm EPS	34.0	9.7	8.5	8.5	4.0	1.2
EIFS, 16 cm EPS	125.8	10.9	31.6	9.7	4.0	1.2
Fast bricks, d = 14 cm, insulated outside						
+ 4.5 cm MW	19.7	8.7	5.3	7.1	3.7	1.6
+16.5 cm MW	75.0	11.5	21.1	10.0	3.7	1.5
Light weight fast bricks, d = 14 cm, insulated outside						
+ 2.5 cm MW	9.6	9.4	3.8	7.0	2.6	2.4
+15.0 cm MW	47.4	12.5	18.5	10.2	2.6	2.3
Fast bricks, d = 19 cm, insulated outside						
+ 4.0 cm MW	27.5	10.2	8.5	8.8	3.6	1.5
+16.0 cm MW	110.7	13.0	31.0	11.6	3.6	1.5
Light weight fast bricks, d = 14 cm, insulated outside						
+ 1.5 cm MW	12.6	11.2	5.0	8.8	2.6	2.4
+14.0 cm MW	76.0	14.4	30.4	12.0	2.5	2.3

9.4 Massive walls with outside insulation

Moisture tolerance

The analysis concerns wind-driven rain and interstitial condensation. Inserting a damp proof course above grade in outer and partition walls prevents rising damp. Building moisture only hampers walls finished vapour-tight at both sides. With thermal transmittances below 0.6 W/(m²·K) and proper ventilation, mould and surface condensation becomes rare, except in the presence of thermal bridges.

Wind-driven rain

Clad stud systems

Usually such systems function as two-step rain control solutions: the cladding as drainage plane, the cavity and/or insulation as capillary break and the massive wall as airtight layer. Solar driven vapour flow is not a problem as most claddings are hardly capillary or, if they are, too thin to buffer enough water to pose problems.

EIFS-systems

Here, the stucco acts as one step rain control layer, combining drainage with some buffering, whereas the insulation must prevent buffered moisture from wetting the massive wall. No rain may penetrate the stucco, let alone allowing it to seep across the insulation. At any rate, insufficient bonding between base layer and decorative stucco could allow water to infiltrate between both and seep across the joints between the insulation boards, reaching the massive wall that way. Cracks in the stucco at insulation board joints also facilitate seepage. Stucco does not buffer much water, 0.6 to 2 l/m² at the maximum. Besides, the capillary water absorption coefficient is usually very low, less than 0.027 kg/(m²·s$^{0.5}$), by which wind-driven rain quickly runs off, increasing the water load on joints around windows and outer doors, facade protrusions and sills. The outcome could be leakage, for example between sills and the insulation/window rebate. A careful study and execution of such details therefore is mandatory.

Stucco that can dry between two wind-driven rain events serves durability. Following requirements are a guarantee:

$$A \leq 0.0083 \text{ kg/(m}^2 \cdot \text{s}^{0.5}) \qquad \mu d \leq 2 \text{ m} \qquad A \mu d \leq 0.0033 \text{ kg/(m} \cdot \text{s)} \qquad (9.18)$$

with A the capillary water absorption coefficient in kg/(m²·s$^{0.5}$) and μ the vapour resistance factor of the stucco. The formulas reflect a simple model, with stucco wetting given by:

$$m_c = w_c \, x = A \sqrt{t_1} \quad (\text{kg/m}^2)$$

with t_1 length of the rain event with a probability of once a year and w_c the capillary moisture content of the stucco. Total drying in turn becomes:

$$m_d = \Delta p \, t_2 / (\mu N d) \quad (\text{kg/m}^2)$$

with t_2 the elapsed time separating two rain events, Δp the vapour pressure difference between the wet stucco and outside and d the stucco thickness. Equating both gives:

$$A \mu d = \frac{\Delta p \, t_2}{N \sqrt{t_1}} \qquad (9.19)$$

The desired capillary water absorption coefficient is found by presuming the moisture front may only reach the insulation at the end of the rain event, or:

$$A = w_c \, d / \sqrt{t_1}$$

For stucco, thickness is ≈ 0.006 cm and capillary moisture content ≈ 150 kg/m³, thus:

$$A = 0.9 / \sqrt{t_1} \quad (t_1 \text{ in seconds})$$

$A = 0.0083$ kg/(m² · s$^{0.5}$) gives $t_1 \approx 1$ dag, a value close to the rain event with a probability of once a year. Still a diffusion thickness limit is needed to exclude interstitial condensation at the stucco's backside. That is 2 m. The result of 0.0033 for the product $A\mu d$ then follows from marking t_1 as 1 day in [9.19] and choosing a representative value for the period t_2 and the vapour pressure difference Δp.

In reality after each rain event, the second drying period succeeds the first, with the total drying time t_2 of the stucco becoming:

$$t_2 = \underbrace{(w_c - w_{cr}) \, d \left[\frac{1}{\beta \sqrt{p}} - \frac{(w_c + w_{cr}) \, d}{3 \, A^2} \right]}_{1^e \text{ drying period}} + \underbrace{\frac{w_{cr} \, d^2 \, \mu \, N}{2 \, \Delta p}}_{2^e \text{ drying period}} \quad (9.20)$$

Or, the capillary water absorption coefficient co-steers drying. When it is so low the first drying period becomes marginal, the second takes over. With the stucco critically moist, critical moisture content (w_{cr}) equals m_d/d with d stucco thickness in m and m_d total drying in kg/m²:

$$m_d = 2 \, \Delta p \, t_2 / (\mu \, N \, d)$$

Equating with total wetting (m_c) gives:

$$A \, \mu \, d = 2 \, \Delta p \, t_2 / \left(N \sqrt{t_1} \right) \quad (9.21)$$

That relation differs by a factor 2 from [9.19]. It applies to stuccos with short first drying period. Of course, as the product $A \mu d$ is double the value of [9.19], the diffusion thickness can also be doubled:

$$A \leq 0.0083 \text{ kg/(m}^2 \cdot \text{s}^{0.5}) \qquad \mu d \leq 4 \text{ m} \qquad A \, \mu d \leq 0.0066 \text{ kg/(m} \cdot \text{s)} \quad (9.22)$$

The best for drying is top stucco with a lower diffusion resistance factor than the base layer. If not, solar radiation may induce blistering. Vapour pressure in a sun radiated wet stucco in fact may reach 16 000 Pa or more, sufficient for delaminating if bonding between base and top is bad. The same mechanism may induce stucco blistering when applied on vapour tight thermal insulation. Blister probability also increases in dark coloured stuccos and stuccos that lose stiffness when turning wet. Stuccos with mineral base and synthetic top layer are therefore more blister-sensitive then synthetic or mineral stucco systems.

Be that as it may, in contrast to inside insulation, outside insulation causes less problems with wind-driven rain, on condition that all details are designed and executed in a way rain control is guaranteed all over the facade. With EIFS, the base and top stucco layers must be compatible.

9.4 Massive walls with outside insulation

Interstitial condensation

Airtight massive walls insulated at the outside may figure as examples of correct application of the basic rule to avoid deposit: thermally best insulating and most vapour permeable layer at the winter cold side, thermally least insulating but most vapour-retarding layer at the warm side. Even with a vapour retarding insulation, condensation risk remains fictitious in moderate climates. The answer to the question 'how does one get interstitial condensation?' is to use an outside finish as air and vapour tight as possible (Figure 9.33).

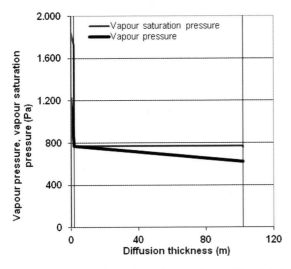

Figure 9.33. Massive wall insulated at the outside, cladding with vapour thickness of 100 m, interstitial condensation for January conditions in a moderate climate.

Clad stud systems

Claddings, made of vapour tight materials, with hardly any or, if many, only airtight joints, give an air and vapour tight finish. While slated claddings do not obey these criteria, stainless steel, copper, aluminium or titanium claddings may. They have infinite vapour resistance and application as storey-high panels with closed joints is not exceptional. Is interstitial condensation a problem in such cases? Sometimes. No, if the moisture deposited at the backside runs off unhindered, is collected, and drained away at the bottom, yes if bolts and butt joints collect run-off and start corroding. The solution must come from an excellent air-tightness of the massive wall, an excellent vapour retarder between massive wall and insulation and/or a ventilated cavity behind the cladding combined with a wind-tight foil covering the insulation.

However, test-building measurements in a moderate climate region on massive walls, insulated at the outside and clad with thin, capillary active fibre cement sheets showed that cavity ventilation hardly affects moisture response. At the rain and sunny south-west side, moisture content in the cladding was almost the same independent of ventilation (Figure 9.34). Where differences appeared, variation in wind driven rain load prevailed. To the northeast, no ventilation effect was noted, but better air-tightness of the massive wall offered an advantage (Figure 9.35).

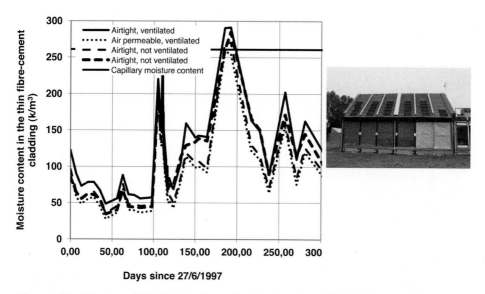

Figure 9.34. Massive wall looking southwest, insulated outside and clad with thin fibre cement sheets. Moisture content in the sheets as measured during a one-year exposure to a moderate, wet climate with a climate class 3 situation indoors.

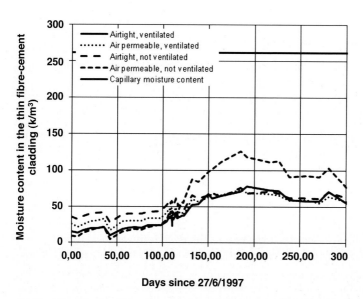

Figure 9.35. Same wall as Figure 9.34, now looking northeast. Moisture content in the sheets.

That cavity ventilation at the sunny side did not comply with expectations was a result of night-time under-cooling, turning the outside air into a moisture source. Additionally, the insulation lacked a wind-tight foil and got air-washed, thereby degrading thermal performance.

9.4 Massive walls with outside insulation

EIFS-systems

Cup tests on 7 mineral and 8 synthetic stuccos gave as diffusion thickness at 86% relative humidity:

Stucco	μd_{86}, m			
	Mean	Standard deviation	maximum	minimum
Mineral	0.32	0.19	0.64	0.09
Synthetic	0.14	0.04	0.19	0.09

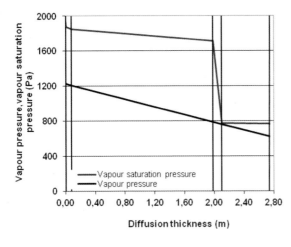

Figure 9.36. Interstitial condensation risk in an EIFS wall, insulated with 10 cm mineral wool, January conditions in a moderate climate.

As shown in Figure 9.36 for a moderate climate January these values are just too low to give interstitial condensation.

Nevertheless, experimental confirmation was judged necessary. For that, a measuring campaign on four one-brick walls, two insulated outside with 5 cm EPS and two insulated outside with 5 cm mineral wool, was initiated. Per couple, the insulation of one wall was finished with synthetic stucco and the insulation of the other with mineral stucco. To compare, the test also included a one-brick wall, insulated at the inside with 5 cm mineral wool and finished with gypsum board. The test took 100 days in a hot box/cold box rig in winter outdoor (2.5 °C, 600 Pa) and climate class 5 indoor conditions (21.5 °C, 2050 Pa). Figure 9.38 summarizes the results.

The four EIFS-walls hardly saw their weight increasing, and one even decreased. Moisture content in the stucco, the insulation and the brickwork after the test did not exceed hygroscopic equilibrium. Instead, due to interstitial condensation, the wall with inside insulation saw its weight increasing by 7.3 kg/m².

A second measurement, now with the one-brick walls wet, underscored that EIFS does not hinder winter drying, except when the stucco is vapour tight. Drying in fact went on as smoothly as for a one-brick wall insulated at its outside with non-stuccoed mineral wool. Even a vapour tight cladding did not stop drying. Of course drying then combined with condensation in the insulation and at the backside of the cladding (Figure 9.38).

The conclusion is clear. Interstitial condensation is no problem in EIFS-insulated massive walls.

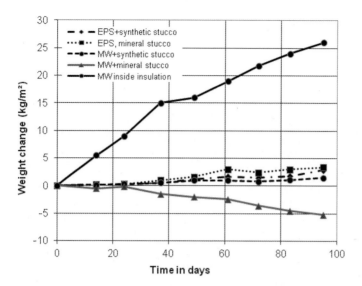

Figure 9.37. Five one-brick masonry walls, one insulated at the inside, the other with EIFS. Moisture uptake or drying under winter outdoor (θ_e = 2.5 °C, p_e = 600 Pa) and humid indoor conditions (θ_i = 21.5 °C, p_i = 2050 Pa).

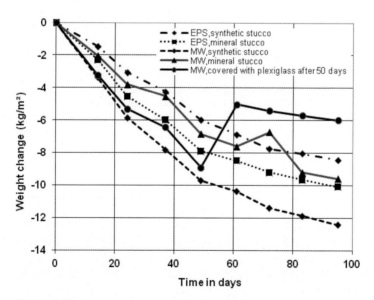

Figure 9.38. Five wet one-brick thick masonry walls, one insulated at the outside with MW, no stucco, the other with EIFS. Drying under winter outdoor (θ_e = 2.5 °C, p_e = 600 Pa) and humid indoor conditions (θ_i = 21.5 °C, p_i = 2050 Pa).

9.4 Massive walls with outside insulation

Thermal bridges

Temperature ratio of an inside partition/massive outer wall connection has been measured experimentally – see Figure 9.39. Where compared to no insulation inside insulation hardly changes surface temperatures in the edge, outside insulation does so in a positive way.

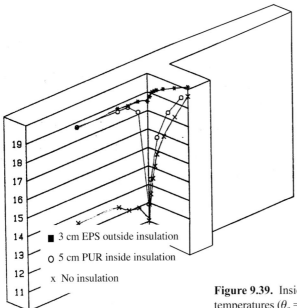

Figure 9.39. Inside wall/massive outer wall connection, temperatures ($\theta_e = 0\ °C$, $\theta_i = 20\ °C$).

Also, the method of detailing the window/outer wall interface was tested by comparing four solutions (Figure 9.40). These are: (1) window behind the outer face of the massive wall, rebate insulated, (2) window behind the outer face of the massive wall, rebate not insulated, (3) window in line with the outer face of the massive wall, outside insulation forming the rebate, (4) window in line with the outer face of the massive wall, no rebate. Clearly, solution (3) performs best: it has the highest temperature ratio. But, even detail (2) has a temperature ratio above 0.7, the value that keeps mould risk below 5%.

Nearly all details discussed in the thermal transmittance section also have temperature ratios above 0.7. Balconies are the only exception, see Table 9.15.

Table 9.15. Details shown in Figure 9.32, temperature ratio (f_{hi}).

Detail	f_{hi}
	–
Lintel (44 cm high)	0.70
Reveal	0.70
Outer sill (wooden inner sill)	0.78
Floor support	0.9
Balcony (without thermal cut between it and the floor slab)	0.63
Roof edge	0.78

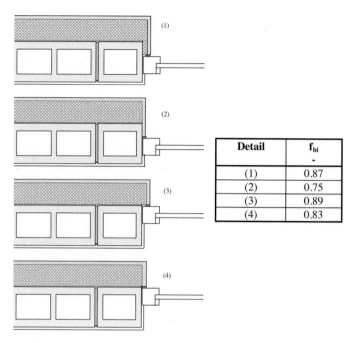

Figure 9.40. Outside insulation, window/outer wall interface.

9.4.2.3 Building physics: acoustics

With EIFS-systems, sound transmission loss drops. The reasons are the thin, lightweight stucco, which induces a high resonance frequency, and the stiff thermal insulation, which decreases damping. Measured data in Table 9.16 confirm the drop. Transmission loss nevertheless remains high enough for the glazed surfaces to fix the envelope's average airborne sound insulation. In addition, in contrast to inside insulation, outside insulation does not increase flanking between terraced dwellings or apartments.

Table 9.16. Average sound transmission loss: massive walls, same wall EIFS-insulated.

Massive wall	R_m dB	Same wall with EIFS	R_m dB
Sand-lime stone, 24 cm, 410 kg/m²	52	60 mm MW, 135 kg/m³	47
		60 mm EPS	46
Lightweight perforated bricks, 24 cm, 210 kg/m²	48	60 mm MW, 135 kg/m³	43
		60 mm EPS	42

9.4.2.4 Durability

Table 9.17 lists the temperatures one could expect in a moderate climate in south-west oriented EIFS-insulated massive walls during cold but sunny winter days and hot, sunny summer days. The massive part sees its temperature hardly change. Through that relative humidity across the part also remains quite constant. The only reason for cracking is mostly initial shrinkage.

The outside cladding, however, is subjected to large temperature changes. Also its relative humidity varies strongly, from 100% when hit by wind-driven rain or suffering from surface condensation by under cooling to 15–20% when sunlit. In addition, claddings react differently from stucco. With slated ones, each unit deforms independently. Problems only arise when slates bend so strongly they push each other away. That may happen with cellulose fibre reinforced cement slates.

With EIFS, the insulation hinders the stucco from deforming. While temperature changes quickly equalize across the stucco's thickness, limiting stress build-up to tension and compression, changes in surface relative humidity cause humidity waves, which cross the stucco slowly while dampened and lagged in time. On a daily basis, the result is the situations shown in Figure 9.41: large variations in relative humidity in the outermost mm, quite stable values at a deeper mm (the words equidistantly and gradually refer to two calculation numerics).

While tension caused by temperature changes usually does not exceed the stucco's tensile strength, bending in the outermost mm induced by a decrease in relative humidity could. Micro cracks are then initiated, which after successive relative humidity waves may become macro cracks. Whether or not this will happen depends on the stucco's tensile strength and deformability, its hygric and thermal expansion ability and the climate. All are stochastic quantities, which is why cracking risk translates in time/failure probabilities, see Figure 9.42. The lower the failure risks within a given period, the more durable the stucco.

Table 9.17. Temperatures in a massive wall with and without EIFS. SW (from row 2 on, stucco relates to the outside surface, 1 is the interface wall/insulation, 2 the inside surface).

Wall		Temperatures, cold winter day		Temperatures, hot summer day	
		Min.	Max.	Min.	Max.
One-brick wall	1	−11.4	7.4	20.9	44.9
	2	8.2	10.8	22.9	31.3
idem, 4,5 cm MW, white stucco	Stucco	−14.4	5.3	19.1	45.7
	1	9.6	11.1	23.4	25.6
	2	15.2	15.4	23.2	23.6
idem, 4,5 cm MW, dark stucco	Stucco	−14.4	25.2	19.2	64.8
	1	10.3	13.4	24.4	28.1
	2	15.4	15.9	23.7	24.6
idem, 16 cm MW, white stucco	Stucco	−15.2	5.0	18.8	46.2
	1	14.4	14.8	23.2	23.8
	2	16.1	16.2	23.1	23.3
idem, 16 cm MW, dark stucco	Stucco	−14.4	25.3	19.1	65.7
	1	14.7	15.5	23.6	24.6
	2	16.2	16.3	23.2	23.4

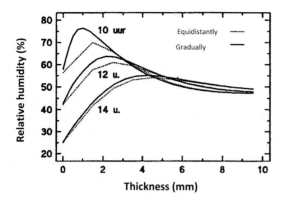

Figure 9.41. Relative humidity waves in 10 mm thick stucco.

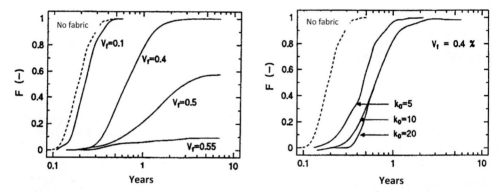

Figure 9.42. EIFS-system. At the left: failure probability for different glass fabric percentages (V_f) as function of time, on the right: failure probability for different adhesive strengths (k_o, MPa) as function of time.

This approach showed stuccos are more crack resistant when:

1. Larger glass fabric fractions are embedded, expressed as a percentage per meter run of stucco. The gain however is not proportional to the percentage. Once above 0.55%, additional improvement in crack resistivity turns to nearly zero.
2. Adhesion between stucco and glass fabric is upgraded. Again, the gain is not proportional to the upgrade. Durability hardly increases once adhesion exceeds 20 MPa/mm.
3. The glass fabric sits closer to the outside surface.
4. The stucco combines higher tensile strength with larger fracture strain.
5. Hygric strain is smaller.
6. Stucco colour is less saturated.

Of the six factors, the first is the most important one followed by: fabric close to the outside surface, good adhesion and small hygric strain. Colour is less important.

9.4 Massive walls with outside insulation

9.4.2.5 Fire safety

Outside insulation should not enhance flame spread. The insulation material is the prime suspect. If flammable and showing fast flame spread, the fire may quickly extend along the envelope to other building compartments. That is why for medium and high rises inflammable insulation of class A1 must be used (mineral wool, cellular glass, mineral foams).

9.4.2.6 Maintenance

With outside insulation, the cladding dictates maintenance intensity. EIFS requires cleaning every 5 to 10 years. For loose dirt, washing suffices. If the dirt sticks to the stucco, brush cleaning with a detergent must be followed by repainting. In choosing paints, declared vapour permeability seems less important. Measurements on site showed vapour permeable paints might cover porous stuccos more effectively than vapour retarding ones. When algae growth is the problem, the stucco must be high pressure water-cleaned and treated with an anti-algae product.

9.4.2.7 Global conclusion

Especially heat, air and moisture checks show that outside insulation out-performs inside insulation. Air-tightness is easier to guarantee, very low whole wall thermal transmittances are within reach, although of less importance transient response is superior, moisture tolerance is much better, not to say quite non-problematic, and thermal bridges are easy to solve. Whenever possible, outside insulation clearly is a preferential choice.

9.4.3 Design and execution

9.4.3.1 Clad stud systems

Detailing and execution here depends on the type of cladding used: timber siding, slated systems, metals (zinc, copper, aluminium, stainless steel, corten steel), natural stone, glass, etc. In all cases, one must design the cladding as a rain barrier. For that, joints between elements do not need caulking. If the massive wall is airtight and a cavity is left between cladding and thermal insulation, pressure equalisation allows open joints. Of course a tray is needed at the bottom of the cavity and wind washing in and behind the insulation must be excluded by covering it with an airtight but vapour permeable spun-bonded foil.

Claddings demand structures to be attached to. For wood siding, timber laths and battens are used. If horizontal, first horizontal battens as thick as the insulation layer are attached to the massive wall. Then, one mounts the insulation boards in between and covers the whole with a spun-bonded foil in overlap. Afterwards, vertical laths are nailed and the siding mounted. Everywhere the facade meets horizontal surfaces, trays fixed behind the spun-bonded foil must assure drainage of the lathed cavity and continuity in rain screening. For metal claddings metal lath and battens are used, preferentially of the same metal as the clad (Figure 9.43), a quite neutral choice being stainless steel. To diminish thermal bridging, perforated profiles or point-wise attachment is recommended. Natural stone claddings finally are hung using anchors that allow correct positioning of each panel while minimizing thermal bridging.

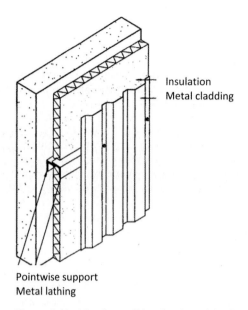

Pointwise support
Metal lathing

Figure 9.43. Massive wall insulated outside, fixing the metal cladding.

9.4.3.2 EIFS-systems

The type of stucco defines how to detail the system. Joints between stucco and other parts anyhow demand careful caulking so run-off rain cannot seep between stucco and insulation and wet the massive wall behind. The execution is as follows. First horizontal guide posts are fixed underneath all outer walls. Then one glues the insulation boards in stretchers bond to the walls, starting at the posts. Gluing is done fully or point wise. If fully, one spreads the bonding mortar over each board's backside with a crest trowel. If point wise, the board's backside gets a strip of bonding mortar along the perimeter and some mortar moulds in the middle. When the wall's outside surface lacks cohesion, one attaches the boards additionally with two or more synthetic nails per board. All insulation boards should have their front-side in the same plane. That allows stuccoing in equal thickness, which reduces the likeliness of cracking. If non-planar, they are levelled. Next, perforated stainless steel corner guides are mounted using bonding mortar, after which one applies the base stucco layer and reinforces it with overlapping glass fibre fabrics. Corners at each door and window bay get extra diagonal strips. If necessary, extra strong fabric is used at street level. The last step consists of applying the top stucco, which must be compatible with the base layer. In order to ensure that, base and finishing stuccos from the same manufacturer should be used (Figure 9.44).

Figure 9.32 showed schematically how to detail joints, roof edges, balconies and others shown in. Window frames are best mounted a little in front of the massive wall. That simplifies lintel and rebate details. To avoid rain seeping behind the insulation, roof edge trims must overlap the EIFS-system. Thermal breaks at balconies are planar with the insulation. Close to grade, the EIFS-system reaches the perimeter insulation, however with a resiliently sealed joint in-between. That way, the stucco will not pick up damp.

If thermal transmittances below 0.15 W/(m^2 · K) are requested, the large insulation thicknesses needed make classic EIFS-systems less desirable. A possibility is an EPS block construction.

9.4 Massive walls with outside insulation

Outer wall to be insulated   Mounting the guide post

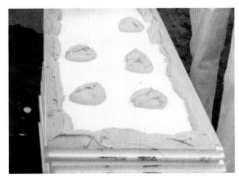   

Pointwise mopping the bonding mortar   Additional fixing with synthetic nails

Fixing the stucco guides   Embedding the glass fibre fabric

Figure 9.44. Massive wall with EIFS, successive execution steps.

First, one builds a building carcass on the foundation walls using the blocks. Then the outer walls are brick-laid and the floor decks cast inside that carcass, its stability being guaranteed by steel strips anchored in the masonry. After one finishes the roof, the EPS-blocks are stuccoed. That allows insulation thicknesses beyond 30 cm, good for clear wall thermal transmittances below 0.12 W/(m$^2 \cdot$ K) (Figure 9.45).

Figure 9.45. Massive walls with EIFS, an alternative way to construct highly insulated assemblies.

9.5 References and literature

[9.1] Anon. (1796). *Enzyclopädie der Bürgerlichen Baukunst* (in German).

[9.2] Gertis, K., Nannen, D. (1985). Thermische Spannungen in Wärmedämmverbundsystemen. Hohlstellen, Versprünge, Rand und Eckbereiche. *Bauphysik 7,* Heft 5, p. 150–157 (in German).

[9.3] Künzel, H. (1985). Regenschutz von Außenwänden durch mineralische Putze. *Der Stukkateur,* Heft 5/85, 8 p. (in German).

[9.4] Knapen, M. (1986). *Massieve buitenwanden met binnenisolatie.* Eindrapport Nationaal Programma RD-Energie, 76 p. (in Dutch).

[9.5] Hens, H. (1986). Licht bouwen, zwaar bouwen? Of, vervangt warmtecapaciteit warmte-isolatie en vice versa. *Tijdschrift Bouwwereld 83,* No. 12, p. 40–43 (in Dutch).

[9.6] Knapen, M., Hens, H. (1986). *Massieve buitenwanden met buitenisolatie.* Eindrapport Nationaal Programma RD-Energie, 86 p. (in Dutch).

[9.7] Gertis, K. (1987). Wärmedämmung Innen oder Außen. *DBZ 35,* Heft 5, p. 631–639 (in German).

[9.8] Hens, H. (1987). *The diffusion resistance of masonry work.* Proceedings CIB-W40 Borås-meeting, 15 p.

[9.9] Laboratorium Bouwfysica. *Invloed van de mechanische bevestiging bij buitenisolatie op de k-waarde van de wand.* Onuitgegeven rapport, 1989, 6 p. (in Dutch).

[9.10] IBBC-TNO. *Trekproeven aan onderdelen van 2 typen buitenisolatie, opgebouwd uit steenwol en afgewerkt met een pleisterlaag.* Rapport 61.6.5375, 1989 (in Dutch).

[9.11] Laboratorium Bouwfysica. *Massieve wandconstructies in cellenbeton.* Rapport 89/3, 1989, 41 p. (in Dutch).

9.5 References and literature

[9.12]   Künzel, H., Leonhardt, H. (1990). *Wärmedämmverbundsysteme mit mineralischen Dämmschichten und Putzsystemen.* IBP Mitteilung 192, 17 (in German).

[9.13]   Lukas, W., Kusterle, W. (1992). *Vermeidung von Putzschäden-Putzinstandsetzung.* Berichtsband der 2. Fachtagung, Innsbruck-Igls, 15–17 Januar, 154 p. (in German).

[9.14]   Carmeliet, J. (1992). *Duurzaamheid van weefselgewapende pleisters voor buitenisolatie.* Doctoraal proefschrift, K. U. Leuven, 202 p. (in Dutch).

[9.15]   Kalksandstein-Information. *Kostengünstiger Wärmeschutz,* 1994, 58 p. (in German).

[9.16]   TI-KVIV. *Studiedag dragend metselwerk.* Proceedings, december 1994 (in Dutch).

[9.17]   Künzel, H. (1995). *Schäden an Fassadenputzen, Schadensfreies Bauen, Band 9.* IRB Verlag, 119 p. (in German).

[9.18]   Hauser, G., Otto, F., Stiegel, H. (1995). *Einfluß von Baustoff und Baukonstruktion auf den Wärmeschutz von Gebäuden, Bundesverband Porenbetonindustrie.* Porenbericht 15, 67 p. (in German).

[9.19]   Van Boxtel, R. (1995). *Handboek gevelisolatie.* Copy Graphic, 258 p. (in Dutch).

[9.20]   TI-KVIV. *Studiedag Massieve wandconstructies.* Proceedings, December 1995 (in Dutch).

[9.21]   Brown, W., Ullett, J., Karagiosis, A. (1997). Barrier EIFS clad walls: results from an moisture engineering study. *Journal of Thermal Insulation and Building Envelopes,* Volume 20, p. 206–226.

[9.22]   Künzel, H. M., Riedl, G., Kießl, K. (1997). Praxisbewährung von Wärmedämmverbundsystemen. *DBZ 45,* Heft 9, p. 97–100 (in German).

[9.23]   Hens, H. (1998). Performance Prediction for Masonry Walls with Inside Insulation Using Calculation Procedures and Laboratory Testing. *Journal of Thermal Envelope & Building Science,* Volume 22, p. 32–48.

[9.24]   Carmeliet, J. (1999). Optimal estimation of damage parameters from localization phenomena in quasi-brittle materials. *Mechanics of cohesive-frictional materials,* Volume 4, p. 1–16.

[9.25]   Stopp, H., Fechner, H., Strangeld, P., Häupl, P. (2001). *The Hygrothermal performance of External Walls with Inside Insulation.* Proceedings of the Performances of Envelopes of Whole Buildings VIII Conference, Clearwater Beach, Florida (CD-ROM).

[9.26]   Hens, H. (2002). Performance Prediction for Masonry Walls with EIFS Using Calculation Procedures and Laboratory Testing. *Journal of Thermal Envelope & Building Science,* Volume 25, p. 167–186.

[9.27]   Künzel, H. M., Künzel, H., Holm, A. (2004). *Rain Protection of Stucco Facades.* Proceedings of the Performances of Envelopes of Whole Buildings IX Conference, Clearwater Beach, Florida (CD-ROM).

[9.28]   Kvande, T., Bergheim, E., Thue, J. V. (2008). *Durability of External Thermal Insulation Composite Systems with Rendering.* Proceedings of the Building Physics Symposium, Leuven, October 29–31.

[9.29]   Krus, M., Hofbauer, W., Lengsfeld, K. (2008). *Microbial Growth on ETICS as a Result of New Building Technology.* Proceedings of the Building Physics Symposium, Leuven, October 29–31.

[9.30]   Buxbaum, C., Pankratz, O. (2008). *Drying Performance of Masonry Walls with Inside Insulation Exposed to Different Exterior Climate Conditions.* Proceedings of the Building Physics Symposium, Leuven, October 29–31.

[9.31]   Künzel, H. M., Zirkelbach, D. (2008). *Influence of Rain Leakage on the Hygrothermal Performance of Exterior Insulation Systems.* Proceedings of the 8th Nordic Symposium on Building Physics, Copenhagen, 16/6–18/6.

[9.32] Barreira, E., Freitas, V. P., Ramos, N. (2009). *Risk of ETICS defacement-As sensitivity Analysis of the Demand Parameters, Energy Efficiency and New Approaches.* Istanbul Technical University, pp. 317–324.

[9.33] Mensinga, P., Straube, J., Schumacher, C. (2010). *Assessing the Freeze-Thaw Resistance of Clay Brick for Interior Insulation Retrofit Projects.* Proceedings of the Performances of Envelopes of Whole Buildings XI Conference, Clearwater Beach, Florida (CD-ROM).

[9.34] Barreira, E., de Freitas, V. P. (2011). *Hygrothermal Behaviour of ETICS-Numerical and Experimental Study.* Proceedings of the 9th Nordic Symposium on Building Physics, Tampere, 29/5–2/6.

[9.35] Vereecken, E., Roels, S. (2011). *A Numerical Study of the Hygrothermal Performance of Capillary Active Interior Insulation Systems.* Proceedings of the 9th Nordic Symposium on Building Physics, Tampere, 29/5–2/6.

10 Cavity walls

10.1 In general

In rainy climates, errors in bricklaying may cause seepage across massive walls. In the early twentieth century, the idea so emerged to split outer walls in a brick veneer and an inside leaf with an air layer in between, called the cavity.

In a cavity wall, the outside surface figures as main drainage plane, the veneer as rain buffer and the veneer's cavity side as a second drainage plane, while the cavity acts as a capillary break preventing cavity side run-off from reaching the inside leaf, which in turn ensures airtightness and load-bearing capacity. Where the cavity closes horizontally, a tray has to guide the cavity run-off back to the outside across open head joints in the veneer. That functional split between veneer and inside leaf was not always clear. In the 1940s, cavity walls with load-bearing one-brick veneer and non-bearing half-brick inside leaf were still being built.

To stabilize the half-brick veneer during brick-laying, wall ties are used to couple it to the inside leaf. A drip nose prevents these from acting as water paths. Above each door and window bay, both the veneer and the inside leaf demand underpinning. Earlier, the veneer figured as outer face formwork for the lintel needed (Figure 10.1).

Figure 10.1. From massive wall to cavity wall with cavity trays and wall ties.

Cavity walls are common in the UK, the Netherlands, Northern Germany, Denmark, the southern part of Sweden, Northern France and Belgium. There are subtle differences in assembly depending on the country:

Country	Veneer wall	Inside leaf
Belgium	9	14
UK	9	9
The Netherlands	11.5	11.5

In Belgium, the inside leaf mostly consists of fast bricks, $L \times B \times H = 29 \times 14 \times 14$ cm^3. Alternatives are sand lime blocks $L \times B \times H = 100 \times 60 \times 14$ cm^3 or concrete blocks $L \times B \times H$

= 29 × 14 × 14 cm³. The veneer is half-brick facing masonry $L \times B \times H$ = 19 × 9 × 4.5 cm³ or 19 × 9 × 6.5 cm³. Sporadically facing concrete blocks with thickness of 9 cm are used. Remarkable variants in execution existed between countries. In the Netherlands, Germany and Denmark the inside leaf was brick-laid before the veneer, demanding a building high scaffold outside. In Belgium and the UK both were brick-laid together, with the veneer a little ahead, which allowed working with a one floor high scaffold inside.

Before 1973, everyone felt happy with a clear wall thermal transmittance (U_o) of 1.2 to 1.7 W/(m² · K). In comparison with massive walls, the value hardly depended on weather conditions, an advantage important enough to designate cavity walls as better insulating. With the energy crises of the 1970s, a clear wall thermal transmittance ≤ 0.6 W/(m² · K) became the objective. Logic dictated filling the cavity with insulation. That way, partial and full fill were born, whereas manufacturers of masonry blocks advocated a third way, brick laying the inside leaf with insulating blocks:

- Partial fill
  The insulation layer is tightened against the inside leaf with in-between insulation and veneer in a 3 to 4 cm wide cavity. The insulation boards are kept in place using ties with fixing strips or spacers (Figure 10.2).
- Full fill
  The insulation layer fills the cavity except for a finger space left behind the veneer (Figure 10.2). Provided the right insulation is chosen and correctly mounted, the fill has to take over all cavity functions: pressure equalization, capillary break, avoiding cavity run-off from reaching the inside leaf.
- Insulating blocks
  The inside leaf is brick-laid using lightweight fast bricks, lightweight concrete blocks or cellular concrete blocks (Figure 10.2).

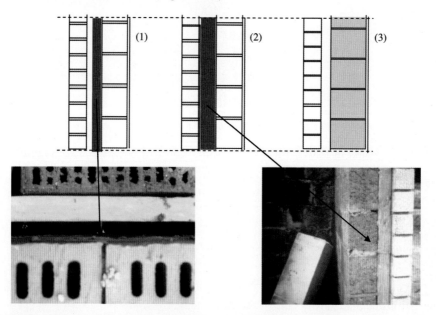

Figure 10.2. Cavity wall: (1) partial fill, (2) full fill, (3) insulating blocks.

10.2 Performance evaluation

In the post-insulation of existing cavity walls, a full fill with injected or blown insulation was the only option left.

Where partial fill and insulating blocks were considered from the start as problem-free, full fills were blasted, mainly due to a misunderstanding on how cavity walls function.

10.2 Performance evaluation

10.2.1 Structural integrity

As the inside leaf is load bearing, it has to carry its own weight plus part of the own weight, dead load and live load of the floor decks, while adding to the building's stiffness against horizontal loads. Dimensioning is done as it is for other load-bearing walls: load summation and controlling stresses, deformation, buckling and fracture safety.

A veneer has to withstand its own weight only. In case of perfect pressure equalization between cavity and outdoors, wind does not add bending loads. Perfect equalization however is never the case. The cavity in fact averages wind pressure, thereby creating local differences with the value outdoors. Accompanying suction and compression are transmitted to the inside leaf by four wall ties per square meter. In medium and high rises, one needs more and stronger ties, whereas compartmentalizing the cavity is good practice there (Figure 10.3).

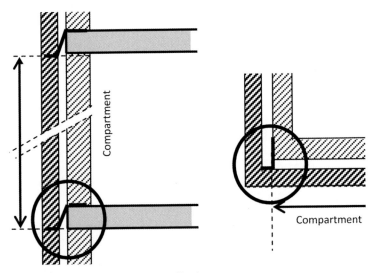

Figure 10.3. Cavity compartmentalization.

Above bays, lintels have to underpin veneer and inside leaf. Lintel loading looks as shown in Figure 10.4: an equilateral triangle of masonry with those in that triangle as floor loads. For serviceability reasons, lintel height is limited to 1/10 of the span. Figure 10.1 shows how concrete lintels were cast in the past. Constructing that way while filling the cavities causes pronounced thermal bridging today. Closing the cavity around door and window bays to assure veneer stiffness and ease of window and door mounting also conflicts with minimizing thermal bridging (Figure 10.4).

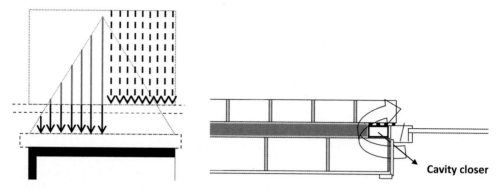

Figure 10.4. Lintels, loading scheme/Cavity closer at a window bay.

10.2.2 Building physics: heat, air, moisture

10.2.2.1 Air tightness

As mentioned, the inside leaf must guarantee air-tightness. Table 10.1 lists a number of measured air permeances. For practice, the table condenses into:

1. Veneer wall, pointed, two open head joints per meter run $\quad K_a = 2 \cdot 10^{-3} \, \Delta Pa^{-0.45}$
2. Inside leaf, pointed, not plastered, air-tight blocks $\quad K_a = 4 \cdot 10^{-5} \, \Delta Pa^{-0.25}$
3. Inside leaf, pointed, not plastered, air-permeable blocks $\quad K_a = 4 \cdot 10^{-4} \, \Delta Pa^{-0.25}$
4. Inside leaf, plastered, air permeable blocks $\quad K_a = 1 \cdot 10^{-5} \, \Delta Pa^{-0.2}$

A veneer is fairly air permeable. This has consequences. Cavity venting and ventilation for example hardly differs between well and no open head joints up and down the veneer, see Figure 10.5.

A non-plastered inside leaf also shows quite a high air permeance, especially when laid with air permeable blocks. In other words, it is the plaster that ensures air tightness. Omitting it, as do some designers, who like fair-face masonry, is irresponsible. Without plaster inside, badly filled head joints in the inside leaf completely kill air tightness. A cavity fill mounted as usual hardly makes a difference. Mineral wool as well as synthetic foam boards without taped joints give air permeances of $2.5 \cdot 10^{-3} \, \Delta Pa^{-0.11}$. Taping the joints between foam boards reduces that value to $7 \cdot 10^{-5} \, \Delta Pa^{-0.11}$, i.e. by a factor of 30. As a consequence, cavity wall air permeances vary between $10^{-6} \, \Delta Pa^{-0.28}$ for the inside leaf plastered and all joints between foam boards taped, and $3.7 \cdot 10^{-4} \, \Delta Pa^{-0.28}$ for a fair face inside leaf and mineral wool or foam boards without taped joints as cavity fill. At an overpressure indoors of 2 Pa, exfiltration varies from 0.005 m³/(m²·h) to 1.8 m³/(m²·h), i.e. × 370!

Wind washing and air looping also demand consideration. Many judge cavity ventilation, which creates an outside airflow in the cavity, positively. Its absence is used as an argument to cover the floor with full fills. Two open head joints per meter run up and down the veneer per story take care of it, though, as Figure 10.5 underlined, cavity ventilation also happens without it.

10.2 Performance evaluation

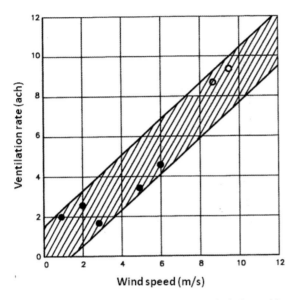

Figure 10.5. Ventilation rate as function of wind speed in a non-ventilated (black dots) and well ventilated (white dots) cavity.

Figure 10.6. Insulation not linking up with the inside leaf (left), not touching the cavity tray (right).

However, when, as is often the case with a partial fill, the insulation layer does not link up with the inside leaf, when open joints are left between the boards, when the layer does not touch the cavity tray and when an open space is left above, then wind washing and air looping will short-circuit the insulation, which is detrimental for the wall's thermal transmittance (Figure 10.6). Sloppy insulation mounting also creates air looping, even without cavity ventilation.

Do air looping and wind washing disappear with an insulating inside leafs? Not when the blocks used are perforated and carelessly laid. Wind washing and air looping may develop across the perforations. Also badly filled head joints get wind washed.

Table 10.1. Air permeance of veneer walls and inside leafs.

Wall	Execution	$K_a = a \Delta P_a^{b-1}$ $(kg/(m^2 \cdot s \cdot Pa))$	
		$a \, (\cdot 10^{-4})$	$b - 1$
Veneer wall, massive facing bricks, 19 × 9 × 4.5 cm	**Head joints carelessly filled**		
	No open head joints	1.22	−0.26
	1 open head joint per m run	3.10	−0.33
	2 open head joints per m run	6.46	−0.38
	3 open head joints per m run	10.64	−0.38
	Joints pointed		
	No open head joints	0.35	−0.19
	1 open head joint per m run	4.83	−0.43
	2 open head joints per m run	9.87	−0.43
	3 open head joints per m run	12.05	−0.43
Veneer wall, perforated facing bricks, 19 × 9 × 6.5 cm	**Head joints carelessly filled**		
	No open head joints	7.48	−0.32
	1 open head joint per m run	21.70	−0.36
	2 open head joints per m run	51.38	−0.48
	3 open head joints per m run	60.98	−0.44
	Head joints well filled		
	No open head joints	1.59	−0.29
	1 open head joint per m run	17.10	−0.47
	2 open head joints per m run	45.30	−0.50
	3 open head joints per m run	72.60	−0.56
	Joints pointed		
	No open head joints	0.39	−0.19
	1 open head joint per m run	16.80	−0.45
	2 open head joints per m run	35.00	−0.44
	3 open head joints per m run	60.45	−0.49
Inside leaf, fast bricks, 29 × 14 × 14 cm	**Head joints carelessly filled**		
	Not plastered	27.50	−0.41
	Head joints well filled		
	Not plastered	0.23	−0.21
	Plastered	0.11	−0.22

Table 10.1. (continued)

Wall	**Execution**	$K_a = a\ \Delta P_a^{b-1}$ (kg/(m² · s · Pa))	
		$a\ (\cdot 10^{-4})$	$b - 1$
Inside leaf, perforated concrete blocks, 29 × 14 × 14 cm	**Head joints carelessly filled**	40.30	–0.42
	Not plastered	40.30	–0.42
	Head joints well filled		
	Not plastered	3.99	–0.27
	Plastered	0.12	–0.23
Inside leaf, glued cellular concrete blocks, 60 × 24 × 14 cm	**Head joints carelessly filled**		
	Not plastered	2.39	–0.39
	Head joints well filled		
	Not plastered	0.99	–0.36
	Plastered	0.13	–0.24
Inside leaf, massive no-fines concrete blocks, 19 × 9 × 9 cm	**Head joints well filled**		
	Not plastered	1.71	–0.13
Inside leaf, massive concrete blocks, 19 × 9 × 9 cm	**Head joints well filled**		
	Not plastered	2.43	–0.14
Inside leaf, perforated light weight no-fines concrete blocks, 19 × 9 × 9 cm	**Head joints well filled**		
	Not plastered	6.16	–0.30
Inside leaf, perforated light weight concrete blocks, 19 × 9 × 9 cm	**Head joints well filled**		
	Not plastered	4.16	–0.25

10.2.2.2 Thermal transmittance

Cavity wall airtight, neither wind washing nor air looping

The clear wall thermal transmittance follows from the well-known heat transmission formula. A value of ≤ 0.6 W/(m² · K) is an absolute upper limit today. The optimum in terms of life cycle costs is around 0.2 W/(m² · K), while passive buildings demand values reaching 0.1 W/(m² · K). Table 10.2 gives the partial fill thicknesses needed, Table 10.3 the full fill thicknesses needed for insulation materials that tested positively on applicability. Both tables underscore that extremely low clear wall values are no evidence. The passive building reference for example demands a glued 14 cm thick inside leaf in light-weight fast bricks and, if PUR is used as partial fill, a 23 cm wide cavity, which together with the 9 cm thick veneer and 1 cm inside

Table 10.2. Partial fill, insulation thicknesses.

Inside leaf	U_o-value W/(m² · K)	Insulation thickness cm					Cavity width ≥ 4 cm
		MW	PUR	EPS	XPS	CG	
Fast bricks, 14 cm	0.6	3.5	2.0	3.5	3.0	4.0	6–8
	0.4	6.5	4.0	7.0	6.0	7.5	8 to 12
	0.2	15.5	10.0	16.5	14.0	18.0	14–22
	0.1	33.5	21.5	35.5	30.0	39.0	26–43
Light-weight fast bricks, glued, 14 cm	0.6	1.5	1.0	1.5	1.0	1.5	6–7
	0.4	4.5	3.0	4.5	4.0	5.0	7–9
	0.2	13.5	8.5	14.0	12.0	15.5	13–20
	0.1	31.5	20.0	33.5	28.0	36.5	24–41
Light-weight fast bricks, glued, 19 cm	0.6	0.5	0.0	0.5	0.5	0.5	6
	0.4	3.0	2.0	3.5	3.0	3.5	6–8
	0.2	12.0	7.5	12.5	10.5	14.0	12–18
	0.1	29.5	18.5	31.0	26.0	34.0	23–38

Table 10.3. Full fill, insulation thicknesses.

Inside leaf	U_o-value W/(m² · K)	Insulation thickness cm		Cavity, total width included finger space ≥ 5 cm
		MW	XPS	
Fast bricks 14 cm	0.6	4.0	3.5	5–6
	0.4	7.0	6.5	7.5
	0.2	16.0	14.5	15–16.5
	0.1	34.0	30.5	31–35
Light-weight fast bricks, glued, 14 cm	0.6	2.0	2.0	5–6
	0.4	5.0	4.5	5.5
	0.2	14.0	12.5	13–14.5
	0.1	32.0	28.5	29–33
Light-weight fast bricks, glued, 19 cm	0.6	1.0	1.0	5–6
	0.4	4.0	3.5	7
	0.2	12.5	11.0	12–13
	0.1	30.0	26.5	27–31

10.2 Performance evaluation

plaster give a total wall thickness of 47 cm. Compared to the classic cavity wall thickness of 30 cm, 46 cm means a larger gross floor area for a same net floor area or a smaller net floor area for the same gross floor area. A thicker outside wall also demands thicker foundations walls, adapted sill, reveal and lintel details at window and door bays, a larger roof surface if the net floor area is kept constant, longer rain gutters, etc. When limiting cavity width to 11 cm and wall thickness to 36 cm the following clear wall thermal transmittances are achievable with glued 14 cm thick lightweight fast bricks or glued 14 cm thick cellular concrete blocks inside leaf and 10 cm mineral wool full, or 7 cm PUR partial fill:

Partial fill (PUR) $U = 0.23$ W/(m² · K)
Full fill (MW) $U = 0.25$ W/(m² · K)

For lightweight block leafs the clear wall thermal transmittances of Table 10.4 are achievable. Performance is disappointing as even the best insulating blocks give values above 0.5 W/(m² · K) for 'normal' wall thickness.

Table 10.4. Lightweight inside leaf: clear wall thermal transmittance.

Inside leaf	Value W/(m² · K)		
	Thickness (cm) =		
	14	19	24
Light weight fast bricks	0.78	0.63	0.53
Cellular concrete	0.74	0.60	0.51
Wall thickness	29	34	39

Cavity wall airtight, wind washing and air looping

Wind washing and air looping are a problem with partial fills. The conditions when that occurs were mentioned above. Wind washing receives extra impulses from two open head joints per meter run up and down the veneer, whereas air looping may also develop without open head joints.

None or only open head joints down the veneer wall

The veneer wall is assumed airtight. Thermal stack between the warm air layer at the inside and the cold air layer at the outside of the cavity fill then acts as the driving force for air looping. If the flow is laminar and the inside leaf is also >airtight, then looping flow approximately becomes (Figure 10.7, local resistances too small to be considered):

$$G_a = \left[\sum_{i=1}^{4}(1/K_{ai})\right]^{-1} \Delta p_T = \left\{\frac{24\,\eta_a}{\rho_a}\left[h\left(\frac{1}{d_1^3}+\frac{1}{d_2^3}\right)+\left[d_{ins}+(d_1+d_2)/2\right]\left(\frac{1}{b_1^3}+\frac{1}{b_2^3}\right)\right]\right\}^{-1} \Delta p_T$$

where d_1 and d_2 are the air layer widths at both side of the fill, b_1 and b_2 the leak widths above and below the fill, h the height of the cavity, d_{ins} the insulation thickness and Δp_T thermal stack, given by:

$$\Delta p_T = 0.043\,h\,(\theta_{c2m} - \theta_{c1m}) \qquad (10.1)$$

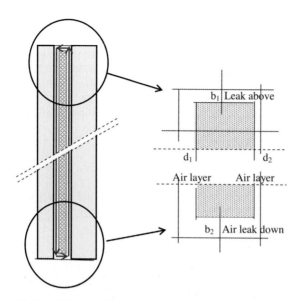

Figure 10.7. Conditions for air looping: air layer at both sides of the fill, leaks up and down or in between the boards.

$\theta_{c1\,m}$ and $\theta_{c2\,m}$ being the harmonic mean temperatures in both air layers. The thermal balance of the two air layers in turn is (radiation and convection combined):

$$\text{Air layer 1} \quad P_{ec1}(\theta_e - \theta_{c1}) + P_{c1c2}(\theta_{c2} - \theta_{c1}) = 1000\, G_a \frac{d\theta_{c1}}{dz}$$

$$\text{Air layer 2} \quad P_{c1c2}(\theta_{c1} - \theta_{c2}) + P_{c2i}(\theta_i - \theta_{c2}) = 1000\, G_a \frac{d\theta_{c2}}{dz}$$

(10.2)

In both equations, P_{ec1} represents thermal permeance between air layer 1 and the outside, P_{c1c2} thermal permeance between both air layers, P_{c2i} thermal permeance between air layer 2 and the inside, θ_{c1} temperature in air layer 1 and θ_{c2} temperature in air layer 2. That the air in both flows in opposition, is considered by pointing the two's z-axis in the flow direction. The system of partial equations rewrites as:

Air layer 1 \qquad\qquad\qquad\qquad Air layer 2

$$\frac{d^2\theta_{c1}}{dz^2} + B\frac{d\theta_{c1}}{dz} + C\,\theta_{c1} = D_1 \qquad\qquad \frac{d^2\theta_{c2}}{dz^2} + B\frac{d\theta_{c2}}{dz} + C\,\theta_{c2} = D_2$$

with: \qquad\qquad\qquad\qquad\qquad\qquad with:

$$B = \frac{P_{ec1} + 2P_{c1c2} + P_{c2i}}{1000\, G_a} \qquad\qquad C = \frac{P_{ec1}\,P_{c1c2} + P_{c2i}\,P_{c1c2} + P_{ec1}\,P_{c2i}}{1000^2\, G_a^2}$$

$$D_1 = \frac{P_{ec1}(P_{c1c2} + P_{c2i})\,\theta_e + P_{c1c2}\,P_{c2i}\,\theta_i}{1000^2\, G_a^2} \qquad D_2 = \frac{P_{c2i}(P_{c1c2} + P_{ec1})\,\theta_i + P_{c1c2}\,P_{ec1}\,\theta_e}{1000^2\, G_a^2}$$

10.2 Performance evaluation

Boundary conditions are:

$$z = 0 \rightarrow \theta_{c1}(0) = \theta_{c2}(h) \qquad z = 0 \rightarrow \theta_{c2}(0) = \theta_{c1}(h)$$
$$z = \infty \rightarrow \theta_{c1} = \theta_{c1\infty}, \; d\theta_{c1}/dz = 0 \qquad z = \infty \rightarrow \theta_{c2} = \theta_{c2\infty}, \; d\theta_{c2}/dz = 0$$

The solution is:

$$\theta_{c1} = \frac{D_1}{C} + \left[C_1 \exp(r_1 z) + C_2 \exp(r_2 z) \right] \quad \theta_{c2} = \frac{D_2}{C} + \left[C_1 \exp(r_1 z) + C_2 \exp(r_2 z) \right]$$

where D_1/C and D_2/C are the temperature asymptotes in both air layers for an infinite height z, equalling the temperatures without air looping, while r_1 and r_2 are the roots of the quadratic equation $\mathbf{D}^2 + B\,\mathbf{D} + C = 0$:

$$r_1 = -\frac{B}{2} - \frac{1}{2}\sqrt{B^2 - 4C} \qquad r_2 = -\frac{B}{2} + \frac{1}{2}\sqrt{B^2 - 4C}$$

The boundary conditions give as integration constants:

$$C_1 = \theta_c(0) - \theta_c(\infty), \quad C_2 = 0$$

The specific solution then becomes:

$$\begin{aligned}\theta_{c1} &= \theta_{c1}(\infty) + \left[\theta_{c1}(0) - \theta_{c1}(\infty)\right] \exp(r_1 z) \\ \theta_{c2} &= \theta_{c2}(\infty) + \left[\theta_{c2}(0) - \theta_{c2}(\infty)\right] \exp(r_1 z)\end{aligned} \qquad (10.3)$$

Temperatures $\theta_{c1}(0)$ and $\theta_{c2}(0)$ follow from ($a = \exp(r_1 h)$):

$$\begin{cases} a\,\theta_{c1}(0) - \theta_{c2}(0) = (a-1)\,\theta_{c1}(\infty) \\ -\theta_{c1}(0) + a\,\theta_{c2}(0) = (a-1)\,\theta_{c2}(\infty) \end{cases}$$

or:

$$\theta_{c1}(0) = \frac{\theta_{c2}(\infty) + a\,\theta_{c1}(\infty)}{1+a} \qquad \theta_{c2}(0) = \frac{\theta_{c1}(\infty) + a\,\theta_{c2}(\infty)}{1+a} \qquad (10.4)$$

The effective thermal transmittance finally looks like:

$$U_{\text{eff}} = \frac{\theta_i - \dfrac{1}{h}\int_0^h \theta_{c2}\,dh}{R_{c2i}(\theta_i - \theta_e)} = U_o \left[1 + \left(\frac{a-1}{a+1}\right) \frac{R_{c1c2}}{r_1\,h\,R_{c2i}} \right] \qquad (10.5)$$

Clearly, the difference between clear wall and effective thermal transmittance increases and thermal quality drops with air looping intensity (r_1 smaller), lower thermal resistance of the inside leaf (R_{c2i} lower) and larger insulation thickness (R_{c1c2} higher). Especially the last is annoying considering the very low clear wall thermal transmittances mandated today.

Air looping intensifies with thermal stack. Larger temperature differences between in- and outdoors, a higher cavity and thicker insulation ensure that. Thermal stack in fact increases proportionally with temperature difference across the fill or, the colder the weather, the worse the insulation quality of carelessly partially filled cavity walls! Height dependence is more complex as the larger flow resistance neutralizes increased stack while the length over which temperature nears the clear wall thermal transmittance enlarges. That pushes the effective value in the direction of the clear wall value.

The impact of air looping has been evaluated by calculation and experiment. Table 10.5 gives measured values, Figure 10.8 the calculated ones. The numerical model behind the curves shown in that figure is far more complex than the one explained. The message is clear: an air layer with a thickness of 10 mm or more between fill and inside leaf and leaks up and down the insulation layer give an up to 180% increase in effective thermal transmittance compared to the clear wall one. The effect of leaks up and down is less straightforward. If between 2 and 4 mm, the ratio between effective and clear wall increases sharply. Once beyond 5 mm things hardly change any more. With an air permeable fill looping losses persist, even with perfect workmanship.

Table 10.5. Partially filled cavity wall, impact of air looping on effective thermal transmittance, hotbox measurements.

Wall properties				Boundary conditions		Thermal properties		
Air layers		Leak width		θ_e	θ_i	U_o	U_{eff}	η_{ins} [1]
Cold mm	Warm mm	Down mm	Up mm	°C	°C	W/(m²·K)	W/(m²·K)	%
Wall 2 m high, cavity 10 cm, 50 mm XPS								
45	5	0	0	1.0	19.7	0.35	0.35	100
		2	2	1.0	20.1	0.35	0.39	89.7
		5	5	0.9	19.1	0.35	0.41	85.4
		18	18	0.9	19.9	0.35	0.42	83.3
40	10	0	0	1.5	21.3	0.34	0.34	100
		2	3	0.8	20.0	0.34	0.49	69.4
		7	3	0.8	19.1	0.34	0.51	66.6
		11	8	1.6	22.4	0.34	0.73	46.6
		22	17	1.6	23.0	0.34	0.75	45.3
25	25	0	0	1.2	23.0	0.35	0.36	97.2
		2	3	1.0	20.8	0.35	0.41	85.4
		5	5	1.0	18.7	0.35	0.73	47.9
		18	18	1.1	18.1	0.35	0.84	41.7
Wall 2 m high, cavity 10 cm, 5 cm MW (20 kg/m³)								
40	10	0	0	0.9	19.7	0.39	0.48	81.3
		13	13	1.1	18.7	0.39	0.66	59.1
30	20	0	0	1.1	22.6	0.39	0.44	88.6
		10	90	1.1	17.1	0.39	0.88	44.3

[1] η_{ins}: insulation efficiency, given by the ratio 100 U_o/U_{eff}

10.2 Performance evaluation

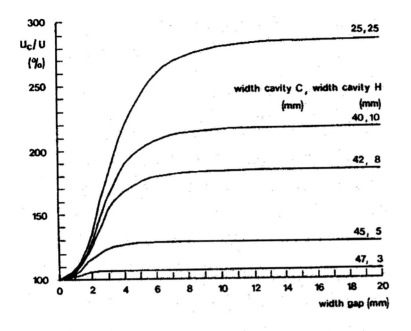

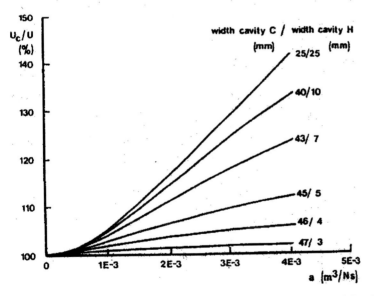

Figure 10.8. Partial fill, with the increase calculated as a percentage of the effective thermal transmittance by air looping compared to the clear wall value, up: for an airtight insulation with leaks up and down the cavity, down: for an air permeable insulation without leaks up and down the cavity.

Preventing the phenomenon demands fills that easily link up with the inside leaf or allow easy sealing of all joints. For glass fibre and mineral wool, density requirements prevail.

Open head joints up and down the veneer wall

Again, the assumptions are that the cavity fill does not link up with the inside leaf, leaks are present up and down the fill and the veneer and inside leaf masonry are assumed airtight. Besides air looping, wind washing also now develops. The hydraulic network looks as sketched in Figure 10.9. For the air permeances of the open head joints per meter run (K_{a5} and K_{a6} in the figure) one may write:

$$K_a = 0.894 \, A_o \, \Delta P_a^{-0.5} \tag{10.6}$$

with A_o their section per meter run. If only wind washing intervened, the outside air should distribute proportionally to the third power of width of the air layers at both sides of the fill:

$$G_a = G_{a1} + G_{a2} \qquad G_{a1} = G_a \frac{d_1^3}{d_1^3 + d_2^3} \qquad G_{a2} = G_a \frac{d_2^3}{d_1^3 + d_2^3} \tag{10.7}$$

A difference in width will make a much larger difference in airflow along. If for example the cold side layer represents 2/3 of the cavity width, than flow distribution will be 89% there and 11% in the 1/3 left at the fill's warm side. Of course, in reality, wind washing and air looping overlap, resulting in three flow patterns: one between the cold side air layer and outdoors, a second between the warm side air layer and outdoors and, a third around the fill. Usually the last is most likely.

Air flows across the open head joints, along both air layers and across the leaks up and down the fill follow from solving the hydraulic network equations, where calculating temperatures and effective thermal transmittance is the same as for air looping – see Equations (10.1)–(10.5).

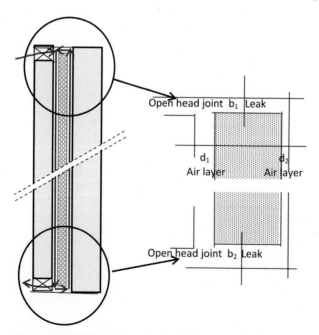

Figure 10.9. Cavity wall with partial fill: hydraulic network with open head joints up and down the (airtight) veneer.

10.2 Performance evaluation

However, when outside air and looping air mix at an open head joint, the mixing temperature becomes the starting value for the air layer washed by both. With pure wind washing, the starting value in both air layers is more or less the outside temperature. As the average temperature in the two depends on flow intensity, iteration between the hydraulic and thermal balances until one reaches a prefixed accuracy is necessary. Table 10.6 lists results for a cavity wall with a fast brick inside leaf, 10 cm XPS as partial fill (total cavity width 15 cm) and the open head joints up and down the veneer 2.5 m away (2 per meter run).

Compared to air looping only, wind washing degrades the clear wall thermal transmittance further.

Table 10.6. Partially filled cavity wall, impact of air looping and wind washing on effective thermal transmittance.

Geometrical wall properties				Boundary conditions			Thermal properties		
Air layers		Leak width		θ_e	θ_i	v_w	U_o	U_{eff}	η_{ins}
Cold mm	Warm mm	Down mm	Up mm	°C		°C	W/(m²·K)	W/(m²·K)	%
Wall 2 m high, cavity 10 cm, 50 mm XPS									
48	2	20	20	0	20	0	0.22	0.224	98.2
				0	20	4	0.22	0.224	98.2
				10	20	0	0.22	0.222	99.1
				10	20	4	0.22	0.222	99.1
45	5	20	20	0	20	0	0.22	0.232	94.8
				0	20	4	0.22	0.288	76.4
				10	20	0	0.22	0.227	96.9
				10	20	4	0.22	0.254	86.6
40	10	20	20	0	20	0	0.22	0.362	60.8
				0	20	4	0.22	0.573	38.4
				10	20	0	0.22	0.295	74.6
				10	20	4	0.22	0.442	49.8
35	15	20	20	0	20	0	0.22	0.561	39.2
				0	20	4	0.22	0.744	29.6
				10	20	0	0.22	0.443	49.7
				10	20	4	0.22	0.647	34.0
25	25	20	20	0	20	0	0.22	0.727	30.3
				0	20	4	0.22	0.832	26.4
				10	20	0	0.22	0.616	35.7
				10	20	4	0.22	0.742	29.6

Cavity wall not airtight, wind washing and air looping

Table 10.7 summarizes the results of a measuring campaign on 12 cavity walls, 6 looking south-west and 6 looking north-east (Figure 10.10), assembled as follows:

1. Capillary facing bricks, 9 cm | Full fill, MW, 14 cm, carelessly mounted | Concrete blocks, 14 cm, quite air permeable
2. Hardly capillary facing bricks, 9 cm | Full fill, MW, 14 cm, correctly mounted | Fast bricks, 14 cm, more air-tight
3. Hardly capillary facing bricks, 9 cm | Partial fill, XPS, 10 cm, carelessly mounted | Concrete blocks, 14 cm, quite air permeable
4. Capillary facing bricks, 9 cm | Partial fill, XPS, 10 cm, correctly mounted | Fast bricks, 14 cm, more air-tight
5. Capillary facing bricks, 9 cm | Full fill, XPS, 10 cm, carelessly mounted | Concrete blocks, 14 cm, quite air permeable
6. Capillary facing bricks, 9 cm | Full fill, XPS, 10 cm, correctly mounted | Fast bricks, 14 cm, more air-tight

The effect of lazy workmanship is alarming: an effective thermal transmittance that is 300% higher than the design value intended. Remarkably, a badly mounted mineral wool full fill shows more stable performance than a badly mounted XPS partial and full fill. After plastering the inside leafs, the situation apparently further degrades. The reason is obvious. As Figure 10.10 shows, higher conduction at the inner face with near zero exfiltration explains the phenomenon.

Table 10.7. Cavity walls, thermal transmittance, design and effective value.

Wall	Cavity fill		U_o-value W/(m²·K)	Measured U-value W/(m²·K)			
				First winter		Second winter	
	Partial	Full		SW	NE	SW	NE
1. −		MF	0.22	0.37	0.32	0.39	0.33
2. +		MF	0.22	0.22	0.21	0.21	0.21
3. −	XPS		0.21	0.86	0.86	0.86	0.86
4. +	XPS		0.21	0.23	0.21	0.23	0.21
5. −		XPS	0.21		0.51	0.60	0.79
6. +		XPS	0.20	0.21		0.22	0.22
				Before air-tightening		After air-tightening	
1. −		MF	0.22	0.39	0.33	0.44	0.35
2. +		MF	0.22	0.21	0.21	0.22	0.22
3. −	XPS		0.21	0.86	0.86	0.94	1.03
4. +	XPS		0.21	0.23	0.21	0.27	0.21
5. −		XPS	0.21	0.60	0.79	0.68	0.94
6. +		XPS	0.20	0.22	0.22	0.22	0.22

− lazy workmanship
+ good workmanship

10.2 Performance evaluation

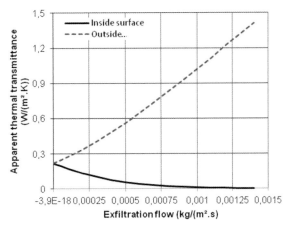

Figure 10.10. At left, the cavity walls tested, on the right an air permeable cavity wall, with air exfiltration impacting the effective thermal transmittance measured at the inside and outside surface.

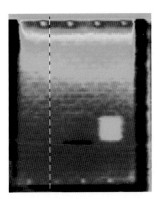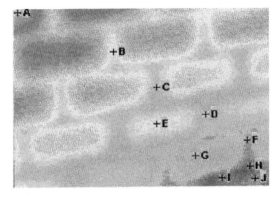

Figure 10.11. Infra red picture of the outside surface of cavity wall 3 and the inside surface of a cavity wall partially filled with stiff insulation boards. Both underscore the existence of air looping (outside warmer up than down, inside colder down than up).

Figure 10.11 shows two infrared pictures: the outer face of cavity wall 3 during a cold winter day and the inner face of a cavity wall partially filled with 3 cm XPS (taken on site). The wall 3 veneer clearly looks warmer above than below, indicating that air looping prevails. On the inner face, mortar joints are not only colder than the fast bricks, both are also colder below than above, proving air looping to be active. That should not happen if the cavity had been correctly filled.

Clearly, for an air permeable cavity wall suffering from air looping and wind washing, thermal transmittance becomes a real unknown. This of course is intolerable, which is why air tightness and exclusion of wind washing and air looping figure are basic requirements when designing and building highly performing cavity walls. The lesson to be learned is that a full fill with mineral wool boards is less risky than a partial fill with stiff insulation boards.

Thermal bridging

Until the early 1980s concrete floors, lintels, columns and beams were cast using the veneer as part of the formwork while cavities were closed around bays (Figure 10.12). That posed no problem because the mean enclosure thermal transmittance (U_m) hardly increased compared to the value based on the clear wall thermal transmittances of the non-insulated cavity walls, on condition the dimensions on the plan and facade were taken out to out, see Table 10.8.

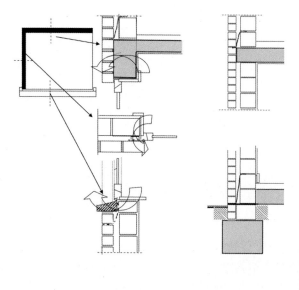

Figure 10.12. Cavity wall detailing as was common practice until the early 1980s. Thanks to the high clear wall thermal transmittance of the unfilled wall, thermal bridge impact remained marginal (Table 10.8).

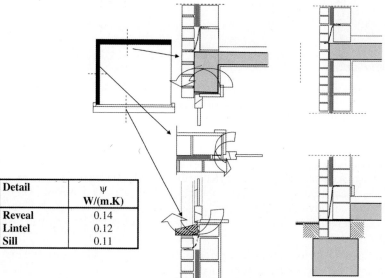

Detail	ψ W/(m.K)
Reveal	0.14
Lintel	0.12
Sill	0.11

Figure 10.13. Filled cavity wall, same details as in Figure 10.12.
Severe thermal bridging now (see linear transmittances at the left and Table 10.8).

10.2 Performance evaluation

Table 10.8. Mean thermal transmittance (U_m) of three dwellings before and after post-filling the cavity (details unchanged).

Case	U_m without TB's W/(m² · K)	U_m with TB's W/(m² · K)	Increase %
Two-family house, 1 floor, area 70 m²			
• Before the retrofit	1.76	1.88	6.8
• After the retrofit	0.54	0.74	37.0
Public housing, 2 floors, total area 70 m²			
• Before the retrofit	1.86	2.05	11.0
• After the retrofit	0.84	1.10	31.0
40 year old end of the row house, 2 floors + loft, area 190 m²			
• Before the retrofit	1.50	1.67	11.3
• After the retrofit	0.62	0.88	41.9

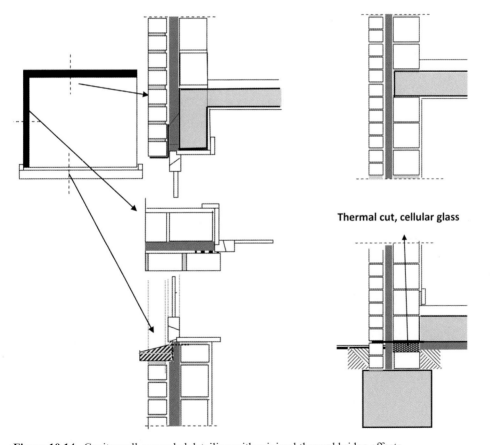

Figure 10.14. Cavity wall, upgraded detailing with minimal thermal bridge effects.

However, after insulation became mandatory things changed. Due to the quite high linear thermal transmittances (ψ) of what now became thermal bridges (Figure 10.13), the mean enclosure thermal transmittance increased drastically (see Table 10.8), which is why filling cavities imposed a redesign of all details. The basic principle was the insulation plane should envelope the whole heated volume without changing plane, a principle demanding thermal cuts where needed and possible. That way the details of Figure 10.14 were developed.

When post-filling existing cavity walls, the thermal bridges existing lintels and cavity closures create may be a true inconvenience, limiting the energy efficiency benefits of the measure and increasing mould probability.

10.2.2.3 Transient response

Under transient conditions airtight partially and fully filled cavity walls perform excellently, as Table 10.9 proves by listing dynamic thermal resistance, temperature damping and admittance of a less and well insulated cavity wall. With light weight inside leaf, admittance limps behind: 2.6 W/(m² · K), lower than the performance threshold of 3.9 W/(m² · K) (for an inside surface coefficient 7.8 W/(m² · K)). For walls where wind washing and air looping have free play, transient quality drops drastically though not for the admittance. The same happens with air permeable walls suffering from infiltration. This does not mean buildings with correctly insulated cavity walls will show excellent temperature damping. That depends largely on glass type, glass area and glass orientation, on solar shading, ventilation strategy and the admittance of floors and inside partitions.

Table 10.9. Filled cavity walls: temperature damping, dynamic thermal resistance, admittance (1-day period).

Wall	Temperature damping		Dynamic thermal resistance		Admittance	
	–	faze, h	m² · K/W	faze, h	W/(m² · K)	faze, h
Partial fill, plastered fast brick inside leaf, 14 cm						
3 cm XPS, $U = 0.56$ W/(m² · K)	23.9	11.3	6.5	9.7	3.7	1.6
12 cm XPS, $U = 0.24$ W/(m² · K)	66.1	12.3	17.8	10.7	3.7	1.6
Full fill, plastered fast brick inside leaf, 14 cm						
5 cm MW, $U = 0.51$ W/(m² · K)	30.4	11.6	8.1	10.0	3.8	1.5
15 cm MW, $U = 0.21$ W/(m² · K)	85.2	12.6	22.6	11.1	3.8	1.5
Partial fill, plastered light weight blocks inside leaf, 14 cm						
3 cm XPS, $U = 0.44$ W/(m² · K)	19.0	12.5	7.4	10.2	2.6	2.4
12 cm XPS, $U = 0.22$ W/(m² · K)	49.4	13.8	19.1	11.5	2.6	1.3
Full fill, plastered light weight blocks inside leaf, 14 cm						
5 cm MW, $U = 0.41$ W/(m² · K)	21.2	12.6	8.2	10.3	2.6	2.3
15 cm MW, $U = 0.19$ W/(m² · K)	56.3	13.8	21.8	11.5	2.6	2.3

10.2 Performance evaluation

10.2.2.4 Moisture tolerance

Building moisture

Building moisture must dry without damage. Part of the initial wetness present in the inside leaf diffuses to the inside. What is left migrates across the cavity fill to the veneer whose backside gets humidified during winter in cold and moderate climates. As soon as the weather becomes warmer, veneer wetness dries mostly to the outside, although solar driven vapour flow may direct some back to the inside leaf. During the first year of building use, inside leaf drying slightly increases end energy consumption for heating because nearly all evaporation heat comes from inside. To allow build-in moisture to dry, painting cavity walls directly after the building is completed cannot be recommended.

Wind-driven rain

Wind-driven rain should humidify neither the cavity fill nor the inside leaf. Also, the veneer may not degrade due to buffering. That insulated cavity walls are by definition rain tight is not a law of the Medes and Persians. During each wind-driven rain event, the facing bricks first suck the impinging raindrops. If the event takes long enough for the outside face to become capillary wet then a water film forms. Once heavier than some 100 g/m², that film runs off, allowing the water to seep through the joints and wet the veneer's cavity side. As soon as the bricks reach capillary moisture content there, seeping turns into cavity side run-off. Measurements show that clinkers with well-pointed joints retard rain seeping and run-off the best, see Table 10.10.

Table 10.10. Rain penetration across an initially dry veneer.

Veneer wall	Air flow across at $\Delta P_a = 10$ Pa m³/(h·m²)	Head joints seeping after ... seconds		Leakage after 1 h	Leakage after 2 h
		First	All		
Capillary facing brick $A = 0.38$ kg/(m²·s$^{1/2}$), $w_c = 300$ kg/m³	20	85"	900"	100%	100%
Idem, carefully pointed	1.5	330"	7200"	30%	100%
Clinker as facing brick $A = 0.07$ kg/(m²·s$^{1/2}$), $w_c = 120$ kg/m³	3.0	80"	9900"	< 10%	80%
Idem, carefully pointed	0.5	1440"	–	< 5%	< 5%

Without a tray inside the cavity, collected run-off may cause rising damp in the inside leaf. Surely in fully filled cavities, mortar drops and wrongly tilted cavity ties could direct run-off into the insulation and to the inside leaf (Figure 10.15). This may be a true problem when post-filling existing cavity walls. Of course, as Figure 10.16 shows, seeping and cavity side run-off is so random that prediction about amounts is almost impossible. Nevertheless, a formula has been proposed in the early 1970s for brick veneers, linking run-off at the cavity side ($G_{r,cav}$) to wind-driven rain intensity (G_{wdr}) and mean wind pressure difference across the veneer (ΔP_a):

$$G_{r,cav} = 2.15 + 0.196 \, G_{wdr} + 0.0308 \, \Delta P_a + 0.0017 \, G_{wdr} \, \Delta P_a \quad (8.34)$$

$$10 \text{ kg/(m}^2 \cdot \text{h)} < G_{wdr} < 40 \text{ kg/(m}^2 \cdot \text{h)}$$

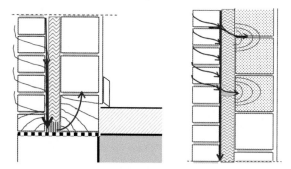

Figure 10.15. Cavity run-off causing rising damp in the inside leaf, ties sloping to the inside leaf allowing seepage to the inside leaf.

Figure 10.16. Concrete block veneer, rain test: seeping joints.

Accordingly, not wind but gravity is the main driving force for run-off. For a wind pressure of zero, run-off in fact remains important. Consequently, head joint height is a defining parameter. The higher it is, the more cavity side run-off can be expected at the rain side. No roof overhang is also negative as it protects the upper part of the outer walls against impinging rain.

10.2 Performance evaluation

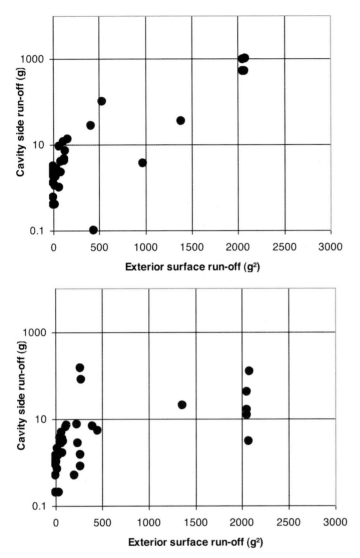

Figure 10.17. Cavity wall with concrete block veneer and cavity fully filled with mineral wool. Up: glued concrete blocks, open head joints; down: mortared concrete blocks. Cavity side run-off compared to outside surface drainage.

During the last two decades, gluing brick veneers with thin-bed mortar gained in popularity. In that case, head joints are typically kept open. One concern is that this practice could increase seeping and run-off. Real world testing since showed that this is not the case, see Figure 10.17. The lower wind pressure differences across the veneer open head joints explained these observations.

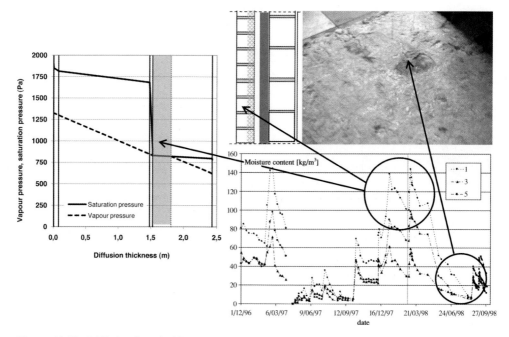

Figure 10.18. Diffusion from inside may prevent a rain buffering veneer to dry completely in winter (left), whereas solar-driven vapour flow from the veneer to inside may cause condensation in a vapour permeable cavity fill and humidify the inside leaf (right).

Winter condensation at their backside and poor drying by under cooling explain the persisting winter wetness rain-buffering veneers of newly and post-insulated cavity walls suffer from in cold and cool climates (Figure 10.18). During the warmer months in such climates, solar driven diffusion from a wet veneer back to the inside causes condensation in vapour permeable fills and moisture deposit against the inside leaf's cavity side. As the heat of evaporation mainly comes from indoors, drying at the start of the next heating season of the humid cavity fill and moist inside leaf heightens the apparent thermal transmittance temporarily by some 10%. Solar driven vapour flow however has a much more pronounced impact in hot and humid climates where indoor cooling and dehumidification are a necessity. Measurements showed that for an outside temperature, 10 °C higher than indoors, a wet veneer succeeds in doubling the clear wall thermal transmittance for a 10 cm thick mineral wool full fill, from 0.29 to 0.58 W/(m² · K). In such climates, one should not use filled brick cavity walls.

With painted veneers that suck water via cracks in the paint but hardly dry or with glazed facing bricks that become wet via the mortar joints, cavity ventilation is thought to enhance drying. Therefore, it should keep temporary thermal transmittance increase acceptable during the next heating season. Many practitioners also believe that ventilation prevents solar driven vapour flow from humidifying a vapour permeable fill and the inside leaf.

As far as drying enhancement is concerned, measurements are less conclusive. Figure 10.19 gives moisture content in the veneer for partially filled ventilated cavity walls and fully filled cavity walls. There is hardly an increase in probability. The same figure also shows the results of drying tests on a fully and partially filled cavity wall after artificially wetting the veneer. The veneer of the partially filled wall dried 2.3 times faster than the one of the fully filled wall.

10.2 Performance evaluation

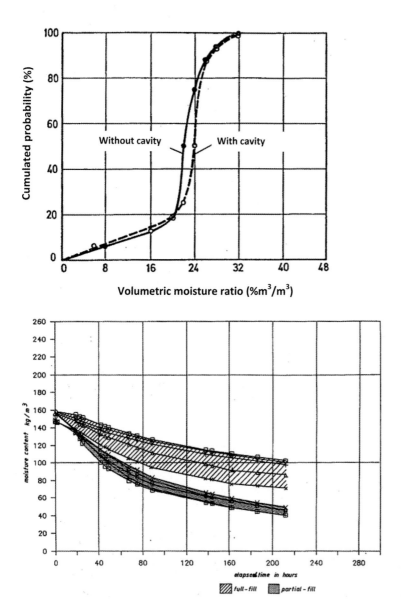

Figure 10.19. Ventilated partially filled and fully filled cavity wall.
Difference in veneer moisture content (up), drying curve in case of a full (upper lines)
and a partial fill, the last with ventilated cavity (lower lines) (down).

This however was not due to a lack of cavity ventilation. In fact, the mineral wool fill contacting the veneer held an important part of the cavity side run-off at its surface. During drying that part diffused back to the veneer's cavity side where it condensed, thereby retarding drying.

Mould and surface condensation

If thermal bridges are absent, if a cavity wall is air-tight and if air looping around and wind washing behind the fill are excluded, then the probability mould will develop and condensate will be deposited against the inside surface drops far below 5%. Even behind cupboards against outside walls, temperature factors easily exceed 0.7:

U_o W/(m² · K)	Temperature ratio behind cupboards ($f_{hi=2}$) –
0.6	0.76
0.2	0.91

The following reasons may play a role when mould complaints nevertheless occur: (1) manifest thermal bridging, (2) air looping and wind washing due to a carelessly mounted partial fill, (3) very tight windows and the lack of a purpose designed ventilation system in the building.

Interstitial condensation

Unpainted veneer, no glazed bricks

- Indoor climate class 1, 2, 3
  If the inside leaf is air-dry and airtight, then moisture deposit at the veneer's cavity side is not a concern – see Figure 10.20 for a northwest facing cavity wall fully filled with 18 cm mineral wool (clear wall thermal transmittance of 0.19 W/(m² · K)). By the end of the winter, the tens of grams deposited critically humidify the facing bricks over some 2.3 mm, a minimal amount compared to the litres of wind-driven rain they buffer.
  Instead, if the inside leaf is air permeable, then the deposit by diffusion and air exfiltration may compete with rain as proven by a hot box-cold box test, comparing veneer humidification by interstitial condensation between an air-tight and air permeable cavity wall – see Figure 10.21. No difference in deposit is seen before an air pressure difference was maintained, while a huge difference appears afterwards. Every winter, veneers of air permeable cavity walls not oriented towards the rain may stay capillary saturated with algae growth in the joints, salt efflorescence and sometimes frost damage as a consequence.
- Indoor climate class 4 and 5
  According to Figure 10.22, even the veneer of a cavity wall with air-tight inside leaf not facing rain will become critically wet, albeit only partially. A vapour retarder at the leaf's cavity side is no option because it is impossible to guarantee continuity, is perforated by cavity ties, etc. Happily, brick laying the veneer with frost-resisting low-salt mortar while using facing bricks of the highest frost resistance class (D) suffices to avoid durability problems.
  Lacking air-tightness of course aggravates the situation. Interstitial condensation will turn the veneer capillary wet and give run-off at its cavity side. That equals the impact rain has. Besides, solar driven vapour flow will create condensation in the cavity fill and moisture deposit against the inside leaf.

10.2 Performance evaluation

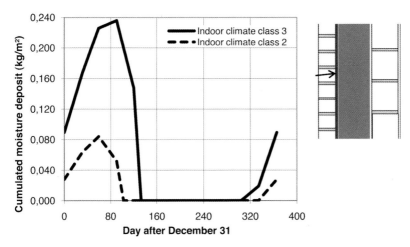

Figure 10.20. Northwest facing fully filled cavity wall with 18 cm mineral wool, $U_o = 0.19$ W/(m² · K), interstitial condensation in indoor climate classes 2 and 3.

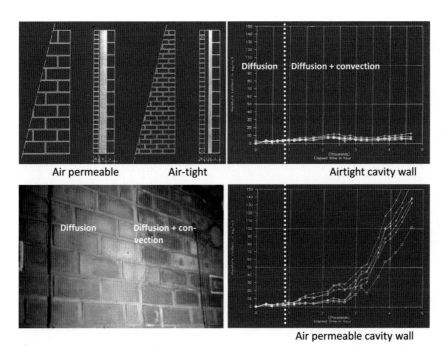

Figure 10.21. Hot box-cold box test on two cavity walls, one airtight, the other air permeable, interstitial condensation by diffusion and diffusion plus convection.

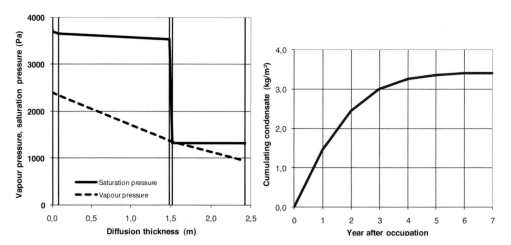

Figure 10.22. Same cavity wall as Figure 10.20, interstitial condensation in indoor climate class 5 (natatorium). On the right the annually accumulating deposit.

Painted veneer or glazed facing bricks

- Indoor climate class 1, 2, 3
  The classic opinion that interstitial condensation was the key problem led to rules such as: (1) do not apply a full fill in such walls, (2) ventilate the cavity through two open head joints per meter run up and down the veneer. As Figure 10.23 shows, a veneer finished with a vapour retarding paint hardly suffers from problematic wetness behind the paint. The small winter deposit there easily dries in summer.
  Still, experience proves painted and glazed veneers are less frost-proof, which means they ultimately become more than capillary wet. In fact, independent of how the cavity is filled, rain uptake by the facing bricks via the mortar joints or micro cracks in the paint is not compensated by equal drying due to the diffusion resistance of the paint or glazed layer reaching values too high for evaporation to be effective. This unbalance allows moisture content in the bricks to accumulate and exceed capillary. Once they become more wet than critical for frost action, damage risk takes off. Avoiding problems therefore demands low-salt facing bricks of the highest frost resistance class (D), brick-laying using low-salt mortar with high frost resistance and high-quality paints. Whether the cavity then is fully filled or not, does not matter. Besides, paint with low vapour resistance factor applied on bricks does not necessarily provide a low diffusion thickness layer. Even so, diffusion resistance still passes the surface film resistance by a factor of 70 to 80.
  For air permeable cavity walls abundant interstitial condensation can make all this more of a problem. Thus, air-tightness must be guaranteed by plastering the inside leaf.
- Indoor climate class 4 and 5
  Accumulating condensate in the veneer may push moisture content beyond critical for frost action, increasing damage risk that way even for no-rain orientations. Painted veneers are also often degraded by salt deposit behind the paint. Lacking air-tightness only aggravates all this. Adding a vapour retarder again is hardly effective, as continuity cannot be guaranteed. The classic recommendations therefore are: (1) apply a partial fill (2) ventilate the cavity. This is no surety, which is why one should not use glazed bricks and painted veneers for indoor climate class 4 and 5 buildings.

10.2 Performance evaluation

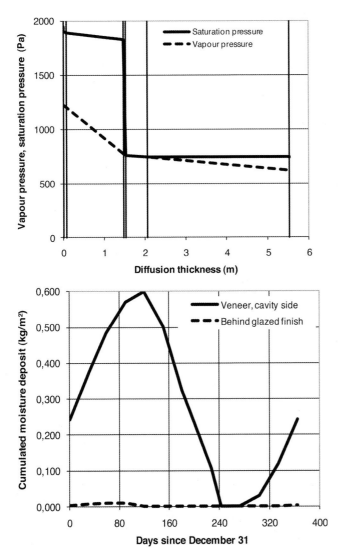

Figure 10.23. Same cavity wall as Figure 10.20, now with glazed facing bricks. Indoor climate class 3, interstitial condensation. Deposit behind the paint is irrelevant.

To summarize for painted veneers or glazed facing bricks:

Indoor climate class	Measures	
1, 2, 3	Cavity wall Veneer	• Airtight (inside leaf plastered inside) • Facing bricks low-salt, frost resistance class D • Mortar low-salt, highly frost resisting • Paint with great covering
4, 5		Do not paint the veneer nor use glazed facing bricks

10.2.2.5 Thermal bridges

As shown in Figure 10.14, all details demand thermal cuts. If absent, the gap between their lowest temperature factor and the temperature factor of the insulated parts increases, not because the inside temperature of all remaining thermal bridges drops, but because the insulated parts become warmer without insulation. Figure 10.24 illustrates this with an infrared picture taken indoors of an outer corner. The isotherms are almost hyperboles.

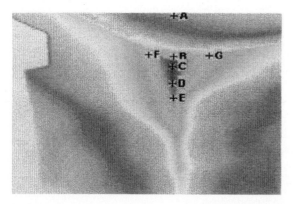

Figure 10.24. Outer corner as geometrical thermal bridge.

Larger temperature differences accelerate local fouling, while temperature ratios below 0.7 raise mould risk above 5%. Table 10.11 gives the lowest temperature ratio for three of the details with correct thermal cut drawn in Figure 10.14.

Table 10.11. Temperature ratio for three details.

Detail	Temperature factor
Reveal	0.79
Lintel	0.79
Sill	0.71

10.2.3 Building physics: acoustics

Airtight filled cavity walls have a sound transmission loss easily exceeding $R_{500} = 52$ dB, though a post-fill with foam contacting both the veneer and the inside leaf can have a negative impact – see Figure 10.25. Even then, as is the case for massive walls, glazed surfaces determine the envelope's sound insulation.

10.2 Performance evaluation

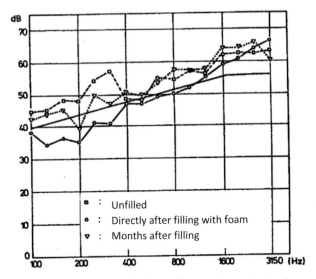

Figure 10.25. Post-filled cavity wall, sound transmission loss.

10.2.4 Durability

Filling the cavity tempers temperature changes in the inside leaf but increases thermal load of the veneer, see Tables 10.12 and 10.13. While the impact of a fill looks marginal in summer, in winter it clearly turns veneers colder with the frost line moving into the fill and subjecting the veneer to more frost-thaw cycles. For example, in a cool climate between December and March 72 cycles were registered at the veneer's outer surface and 44 at its cavity side for an unfilled cavity wall, whereas with filled cavity the numbers were 98 and 86. Frost damage risk increases that way. Moreover, in moderate climate regions, veneers of north facing filled cavity walls show more moss and algae growth than veneers of unfilled walls. Also when micro cracking between brick and mortar increases, so does seepage and cavity side run-off.

Partial fill does not perform better than full fill. Happily, thermal load hardly increases once the insulation thickness passes 3 to 5 cm. Air looping relaxes that load, as the results recorded in Table 10.13 prove. Of course, that little advantage does not compensate for the huge differences between effective and clear wall thermal transmittance.

The following measures are useful: (1) choosing facing bricks of highest frost resistance class, (2) bricklaying the veneer with a smooth mortar (bastard mortar is a possibility, though doubts exist about frost resistance), (3) providing the veneer with expansion joints at regular intervals. The last measure prevails for extended outer walls as well as for terraced housing where expansion joints are advisable at each party wall. If the facade follows a broken line, then all edges absorb the veneer's movement.

Table 10.12. Veneer, calculated temperatures (Uccle, SW, 1 outside surface, 2 cavity side).

Cavity wall		Temperatures, cold winter day		Temperatures, hot summer day	
		minimum	maximum	minimum	maximum
No cavity fill, fast brick inside leaf, 14 cm	1	−13.1	10.6	20.4	49.1
	2	−7.8	4.5	22.1	41.6
Partial fill, fast brick inside leaf, 14 cm, 3 cm XPS	1	−14.3	9.9	20.2	50.0
	2	−11.9	7.5	20.8	46.1
Partial fill, fast brick inside leaf, 14 cm, 12 cm XPS	1	−14.8	9.5	20.0	50.2
	2	−13.4	6.7	20.7	46.8
Full fill, fast brick inside leaf, 14 cm, 5 cm MW	1	−14.4	9.8	20.0	50.1
	2	−12.2	7.4	20.8	46.3
Full fill, fast brick inside leaf, 14 cm, 15 cm MW	1	−14.8	9.5	20.2	50.2
	2	−13.3	6.8	20.7	46.8

Table 10.13. Veneer, measured temperatures, 9 to 30 January 1997 (NE, 1 outside surface, 2 insulation, cavity side).

Cavity wall		Temperatures	
		minimum	maximum
Airtight (fast brick inside leaf, 14 cm, joined and plastered), full fill with 14 cm mineral wool	1	−7.2	10.6
	2	−5.6	10.6
Air permeable (concrete block inside leaf, 14 cm, partial fill with 10 cm XPS, lazy workmanship	1	−5.8	11.1
	2	−1.6	11.6
Airtight (fast brick inside leaf, 14 cm, joined and plastered), partial fill with 10 cm XPS	1	−6.9	11.0
	2	−3.9	11.3
Air permeable (concrete block inside leaf, 14 cm, full fill with 10 cm XPS, lazy workmanship	1	−4.9	11.4
	2	1.7	12.5

10.2.1 Fire safety

A correctly constructed cavity wall has a structural fire resistance easily exceeding 90′. Problems may arise when filling the cavity with a combustible insulation such as EPS and XPS. If the flames then penetrate into the cavity, fire may spread to higher floors.

10.2.1 Maintenance

A facing brick veneer of good quality hardly demands maintenance. On the average cleaning and pointing is only needed every 40 to 50 years.

10.3 Design and execution

10.3.1 New construction

Unfilled cavity walls showed a high tolerance for design mistakes and workmanship errors. Or put otherwise, it was a low risk construction. Filling or post-filling the cavity lowers that tolerance. High quality therefore not only demands correct design but also careful execution with a quality control focussing on all elements essential for the right performance and durability:

1. Airtight
2. No air looping possible around the cavity fill, no wind washing
3. Facing bricks frost resisting, low free salts content
4. Mortar frost resisting, low free salts content
5. Correct cavity trays where needed
6. As few thermal bridges as possible

10.3.1.1 Airtight, as few thermal bridges as possible

Not only must the insulation layer show continuity along the facade, also an airtight finish inside is requested. A plaster layer suffices for that or, if absent, a well-pointed fast brick inside leaf. In case it consists of fair-face no-fines concrete blocks, then the only way to guarantee air-tightness is by rendering the cavity side with a bastard mortar. Air-tightness is not a problem per square meter of wall; difficulties arise around window and door bays, at skirting boards, at plug sockets and switches and where cavity walls contact pitched roofs. Possible solutions are:

- Window and door bays, see Figure 10.26 (inside plaster)
  Caulk the joints between the plastered reveal and the window or doorframe. The alternate consists of plywood boards covering the reveal, the space between plywood and reveal blocks filled with mineral wool or sprayed PUR and all joints with the plaster and window or doorframes caulked. Also caulk the joint between sill and window frame, sill and inside plaster and, sill and plywood boards
- Skirting, see Figure 10.26
  Stop gypsum plastering some 10 cm above screed level. Render these 10 cm with a water repelling mortar. Caulk the remaining joint between skirting and floor finish with resilient putty. These measures also prevent gypsum plaster from sucking cleaning water
- Facade/roof, see Figure 10.26
  Finish the plaster against a steel angle; caulk the joint between angle and roof inside lining, if airtight, with resilient putty. Otherwise, plaster up to 1 cm below the roof's air barrier wall laths and caulk the joint in between with resilient putty.

Continuity of the thermal insulation requires the windows to be mounted in the insulation plane with the veneer as rebate. This is also preferred as rain screening, on condition a waterproof layer is inserted between masonry rebate and window frame.

Lintels are critical in terms of thermal bridging and rain screening. They were and are sometimes still realized using the veneer as formwork, while insulating their back- and underside, see Figure 10.27, left. A makeshift solution, dictated by ease of execution. Thermal bridge effects displace toward the concrete floor while the lintel itself endures large temperature differences with cracking support masonry as a consequence.

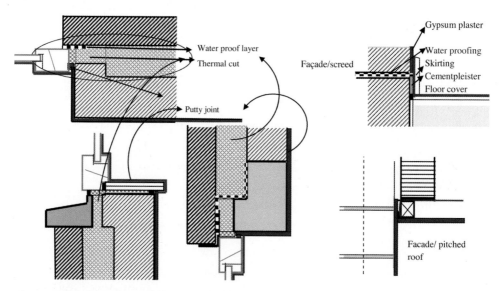

Figure 10.26. Continuity of thermal insulation and air retarding layer inside around window bays, at skirting boards and in the contact facade/roof.

Figure 10.27. Lintels. On the left: how not to insulate, on the right: how to do it.

Correct execution demands mounting the thermal insulation between lintel and veneer (Figure 10.27, right).

10.3.1.2 Correct cavity trays where needed

Cavity trays are needed above grade and above window and door bays. Trays there need side edges to prevent run-off from leaking on the insulation and seeping to the inside leaf (Figure 10.28). Trays above lintels complicate cavity fill execution. As a result, in practice, insulation below the tray is often omitted. The solution? Look where the water runs off: at the cavity side of the veneer. It thus suffices to mount a tray with high enough back flange below that drainage plane. Manufactured lintels, with insulation and a steel L-section underpinning the veneer including the tray, simplify construction.

10.3 Design and execution

Figure 10.28. Cavity trays with side edges.

10.3.1.3 Excluding air looping and wind washing

The following possibilities exist:

- Taping all joints between the insulation boards and caulking the joint above grade between insulation and cavity tray
- Using composite boards consisting of hard front- and soft back layer, so the insulation can be linked up perfectly with the inside leaf
- Gluing the boards with bonding mortar against the inside leaf as is done with EIFS

The measures named have one aspect in common: they all rely on the following execution sequence (Figure 10.29):

- Work with outside scaffolding
- Erect the inside leaf, insert all damp proof barriers needed to form cavity trays at the right location, point the masonry
- Mount windows and doors

Figure 10.29. Correct execution sequence of a high performance filled cavity wall.

- Fix the thermal insulation. Stiff boards are best glued against the inside leaf with bonding mortar, after which the cavity ties are bored. Soft boards or composite boards with soft back layer are linked up with the inside leaf using bored cavity ties. Mount the boards in stretcher bond. Assure continuity at corners by letting out the boards and cutting them to shape
- Brick lay the veneer

That sequence applies as well for a partial as for a full fill, though with full fills one should use bored ties without drip nose that slope away from the inside leaf.

10.3.2 Post-filling existing cavity walls

Specific techniques have been developed to post-fill existing walls. In the 1970s, after the first energy crises, injecting UF-foam was quite popular (Figure 10.30). Shortly after, blown loose fill mineral wool gained applicability, as were injected PUR-foam and poured PS pearls. After 2000, cellular glass granules became available.

Figure 10.30. Post-fill with UF-foam (left),
cavity tray with too low back flange as it looked after removing all mortar drops (right).

UF-foam is no longer used, mainly due to complaints about formaldehyde emission, the severe drying shrinkage the material experiences which lowers thermal performance and the opportunities the two-component application offers to cheat costumers. Measurements in the early 1980s gave moisture ratios up to 382% kg/kg directly after application for a dry density between 7.6 and 16.4 kg/m^3. That ratio dropped to 23% kg/kg half a year after application, causing a volumetric shrinkage between 4.5 and 10%. As the foam has a capillary water absorption coefficient between 0.003 and 0.012 kg/(m$^2 \cdot$ s$^{0.5}$), it could act as capillary bridge between the veneer and the inside leaf. UF is also hygroscopic, very vapour permeable with a vapour resistance factor of 1.4, mould sensitive and temperature sensitive. A hotbox measurement on a cavity wall with assembly (from inside to the outside):

- inside plaster
- inside leaf in 14 cm thick fast bricks
- 6 cm wide cavity post-filled with UF-foam
- 9 cm thick half-brick veneer

gave a measured clear wall thermal transmittance of 0.54 W/(m$^2 \cdot$ K) whereas a value of 0.41 W/(m$^2 \cdot$ K) was calculated using the measured thermal conductivity of the foam, 0.036 W/(m $\cdot$ K).

Things looked as if the foam had an onsite thermal conductivity 0.051 W/(m·K), i.e. 42% higher. The reason was shrinkage of the UF-layer, creating a system of cracks in the insulation and voids at both sides allowing air looping to develop.

Blown loose fill mineral wool, injected PUR with density above 20 kg/m^3, thermal conductivity 0.028 W/(m·K), vapour resistance factor 10, and cellular glass granules took over the market. Injected PUR anyhow wets under water heads and deforms at 70 °C.

To continue, before post-filling an existing cavity wall, a detailed visual and endoscopic inspection is no luxury. One must check the veneer for possible frost damage at unused chimneys and parts separating unheated spaces from outside. If it shows damage, do not fill! Potential thermal bridges (cavity closers around windows, lintels, sometimes massive edges in older cavity walls) have to be looked for. The way trays are mounted and filled with mortar droppings and whether cavity ties bear mortar drops, also requires control. If such evaluation is overlooked, post-filling may initiate unwanted consequences such as mould growth indoors, rain penetration and frost damage. In fact, as was said, a filled cavity wall is less forgiving than an unfilled one.

10.4 References and literature

[10.1]   Uyttenbroeck, J. (1963). Enkele inleidende begrippen betreffende de rol van de spouwmuur. *Tech. Wet. Tijdschrift 32,* No. 6 (in Dutch).

[10.2]   Vos, B. (1963). De thermische en hygrische eigenschappen van de spouwmuur. *Tech. Wet. Tijdschrift 32,* No. 6 (in Dutch).

[10.3]   Künzel, H. (1963). Das Feuchtigkeisverhalten von zweischaligen Wänden und Wänden mit vorgehängten Verkleidungsplatten. *Tech. Wet. Tijdschrift 32,* No. 6 (in German).

[10.4]   Architects Journal, Technical Study, Failures in Brickwork, 6 oct. 1971.

[10.5]   Vos, B. *De spouwmuur, TI-KVIV-cursus Warmteisolatie en vochtproblemen in gebouwen.* Antwerpen, 1976 – 1978 – 1981 – 1983 – 1985 (in Dutch).

[10.6]   Vos, B., Tammes, E. (1976). *Rain penetration through the outer walls of cavity structures.* CIB-W40 meeting, Washington.

[10.7]   WTCB-Informatie. *Termische isolatie en dichtheid van spouwmuren.* Het Bouwbedrijf, 7/1979 (in Dutch).

[10.8]   Vos, B. (1980). *Het isoleren van bestaande woningen door middel van spouwmuurvulling.* Studiedag Termische isolatie, Diegem (in Dutch).

[10.9]   TNO. *Onderzoek naar de regen en condensatieinvloed bij muren, gevuld met polystyreenschuim en minerale wol.* Rapport B80–129, 1980 (in Dutch).

[10.10] Hens, H., Standaert, P. (1981). *Thermal bridges in traditional Belgian masonry houses.* CIB-W40 meeting, Kopenhagen.

[10.11] Siviour, J. (1982). *Thermal performance in practice of cavity walls insulated with urea-formaldehyde foam.* Building Services Engineering Research & Technology.

[10.12] Newman, A., Whiteside, D., Kloss, P., Willis, W. (1982). Full-scale Water Penetration Tests on Twelve Cavity Fills-Part I. Nine Retrofit Fills. *Building and Environment, Vol. 17,* No. 1, p. 175–191.

[10.13] Newman, A., Whiteside, D., Kloss, P., Willis, W. (1982). Full-scale Water Penetration Tests on Twelve Cavity Fills-Part II. Three Built-in Fills. *Building and Environment, Vol. 17,* No. 3, p. 193–207.

[10.14] Wilberforce (1983). *Cavity wall test facilities.* Pilkington Fibreglass, Building technical File 3.

[10.15] Hens, H., Wouters, P. (1983). *Onderzoek in situ naar de kwaliteit van warmte-isolerende ingrepen, Nieuwe perspectieven voor de bouwnijverheid.* Centrum Hoger Onderwijs, Brugge (in German).

[10.16] Künzel, H. (1983). *Wärme- und Regenschutz bei zweischaligem Sichtmauerwerk mit Kerndämmung.* Rapport B Ho 9/83, Fraunhofer Institut für Bauphysik, Holzkirchen, 48 p. + Tab + Fig (in German).

[10.17] Hens, H. (1984). *Buitenwandoplossingen voor de residentiële bouw: de spouwmuur.* Eindrapport Nationaal Programma RD-Energie, 67 p. (in Dutch).

[10.18] Jackson, L. (1985). *Feedback on Cavity Fill.* Pilkington Fibreglass, Building Technical File 9.

[10.19] Knapen, M., Standaert, P. (1985). *Experimental research on thermal bridges in outer wall systems.* CIB-W40 meeting, Holzkirchen.

[10.20] Hoff, W., Platten, A. (1985). *Drying processes in masonry walls: effects on fabric and environment.* CIB-W40 meeting, Holzkirchen.

[10.21] Standaert, P. (1985). *Thermal bridges: a two-dimensional and three-dimensional analysis.* Proceedings of the Thermal Performances of the Exterior Envelopes of Buildings III Conference, Clearwater Beach, Florida.

[10.22] Kreijger, P. (1986). *Aspecten van duurzaamheid en onderhoud van baksteengevels.* 6^e Verbeter mee Symposium, Enschede (in Dutch).

[10.23] Van Es, J., Kreijger, P. (1986). *De invloed van de isolatie in de spouw op het gedrag van spouwmuren.* 6^e Verbeter mee Symposium, Enschede (in Dutch).

[10.24] Laboratorium Bouwfysica. *Spouwmuurisolatie van woningen met minerale wol, deel- en volledige vulling, bouwkundig detail, uitvoering.* Rapport 88/34 (in Dutch).

[10.25] Lecompte, J. (1989). *Influence of natural convection in an insulated cavity on the thermal performance of a wall.* ASTM STP 1030, p. 397–420.

[10.26] Künzel, H. (1989). Keine Probleme bei zweischaligem Mauerwerk mit Kerndämmung. *Baumarkt 89,* Heft 9, p. 631–633 (in German).

[10.27] Vos, B. *Spouwmuurisolatie van woningen met minerale wol.* MWA, 32 p. (in Dutch).

[10.28] Laboratorium Bouwfysica. *Spouwmuurventilatie.* Rapport 90/14, 1990 (in Dutch).

[10.29] Laboratorium Bouwfysica. *The insulated cavity wall: quality of design and workmanship.* Rapport 90/15, 1990.

[10.30] Hens, H., Lecompte, J. (1990). *Filled cavity walls: influence of design and workman-ship on the hygric and thermal performances.* Paper III7, Proceedings of the CIB-W67 Symposium on Energy, Moisture and Climate in Buildings, Rotterdam.

[10.31] Künzel, H. (1991). Wärme- und Feuchteschutz von zweischaligem Mauerwerk mit Kerndämmung. *Bauphysik 13,* Heft 1, p. 1–9 (in German).

[10.32] Region Wallonne, D. G. T. R. *Energie, Isolation thermique et étanchéité d'un mur creux, Situation problèmes.* ULg (in French).

[10.33] Schnaupauff, V., Dahmen, G., Oswald, R. (1993). *Schlagregenschutz von Aussenwänden, zur Bewährung und Beurteilung wassernehmender Fassadenkonstruktionen.* Referat BI 5, Bundesministerium für Raumordnung, Bauwesen und Städtebau (in German).

10.4 References and literature

[10.34] Hens, H., Janssens, A. (1995). *Hygrothermal Response of Filled Cavity Walls*. Proceedings of the International Symposium on Moisture Problems in Building walls, Porto.

[10.35] Hens, H., Mohamed, F. Ali (1995). *Heat-air-moisture design of masonry cavity walls: theoretical and experimental results and practice*. ASHRAE Transactions, Volume 101, part 1, p. 607–626.

[10.36] Mohamed, Ali (1995). *Cavity walls: Extra heat losses due to natural convection*. IEA-Annex 24 on Heat, air and moisture transfer in Insulated Envelopes, Paper T5-B-95/03.

[10.37] Department of the Environment. *Minimising thermal bridging in new dwellings*. Good Practice Guide 174, UK, 1996.

[10.38] Department of the Environment. *Minimising thermal bridging when upgrading existing housing*. Good Practice Guide 183, UK, 1996.

[10.39] Brocken, J. (1998). *Moisture Transport in Brick Masonry: the Grey Area between Bricks*. Doctoral Thesis, TU/e.

[10.40] Hens, H., Janssens, A., Depraetere, W. (2001). *Hygrothermal Performance of Masonry Walls with Very Low U-value: a Test House Evaluation*. Proceedings of the Performances of Envelopes of Whole Buildings VIII Conference, Clearwater Beach, Florida (CD-ROM).

[10.41] Beerepoot, M. (2002). *Energy Regulations for New Buildings, in Search of Harmonisation in the EU*. DUP Science.

[10.42] Van Mook, F. (2002). *Driving rain on building envelopes*. Doctoral Thesis, TU/e.

[10.43] Blocken, B. (2004). *Wind-driven rain on buildings: measurements, numerical modelling and applications*. Doctoral Thesis, K. U. Leuven.

[10.44] Hens, H., Roels, S., Desadeleer, W. (2005). *Glued Concrete Block Veneers with Open Head Joints: Rain Leakage and Hygrothermal Performance*. Proceedings of the 7th Nordic Symposium on Building Physics, Reykjavik, 13/6–15/6.

[10.45] Meeusen, J. (2006). *Na-isolatie van spouwmuren (Post-insulation of cavity walls)*. Master Thesis, UGent.

[10.46] Carmeliet, J., Karagiosis, A., Derome, D. (2007). *Cyclic temperature gradient driven moisture transport in walls with wetted masonry cladding*. Proceedings of the Performances of Envelopes of Whole Buildings X Conference, Clearwater Beach, Florida (CD-ROM).

[10.47] Künzel, H. M., Karagiosis, A., Kehrer, M. (2008). *Assessing the Benefits of Cavity Ventilation by Hygrothermal Simulation*. Proceedings of the Building Physics Symposium, Leuven, October 29–31.

[10.48] Straka, V. (2011). *Water Penetration through Clay Brick Veneer Wall*. Proceedings of the 9th Nordic Symposium on Building Physics, Tampere, 29/5–2/6.

11 Panelized massive outer walls

11.1 In general

Panelized massive outer walls are built using manufactured reinforced concrete elements. Cost factors of such facades are: the number of different elements needed, the lot per element and the form complexity of the elements. Choosing a panel solution only makes sense if the design is modular and detailing such that a limited number of different elements suffices. Transport is done with trucks, meaning the elements cannot be too large or too heavy, and their weight cannot exceed the lifting capacity of the site crane. Once mounted, all joints between elements have to be air- and rain tightened.

Three panel types are in use (see Figure 11.1):

1. Sandwich elements with an outside leaf in facing concrete, an inside leaf in normal concrete and an insulation in between
2. Monolithic elements, after mounting insulated at the inside
3. Monolithic elements, after mounting insulated at the outside

A panelized facade can be load bearing or not. If load bearing, then the panels have to withstand assembling forces, facade weight and part of the own weight, dead weight and the live load of floors. In low-rise construction, the panels must also guarantee wind-stiffness, requiring rigid coupling between panels and floors. In medium and high rises, however, stiff cores do that job. Assembling is done floor-wise (Figure 11.2). Non-bearing panels have to withstand their own weight, wind load and assembling forces. They are mounted after the load bearing structure is finished.

Performances of panels insulated at the inside or the outside are comparable to those of identically insulated massive walls. Sandwich panels instead demand their own solutions. We discuss only these in what follows.

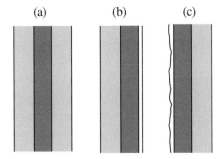

Figure 11.1. (a) Sandwich elements, (b) monolithic elements, insulated inside, (c) monolithic elements, insulated outside.

Figure 11.2. Manufactured load-bearing panels.

11.2 Performance evaluation

11.2.1 Structural integrity

Loadbearing panels have to fulfil the same performance requirements as any loadbearing wall. Non-bearing panels instead are attached storey-wise to the load-bearing structure. In most cases front beams act as supports while being coupled sideway to the front columns or front beams above. In doing so, the supports undergo bending and shear, whereas the strap anchors experience tension. During site assembling, each panel is carefully positioned horizontally and vertically using adjusting screws below and the strap anchors above. Zigzagging is avoided by correctly aligning each panel (Figure 11.3).

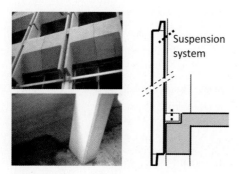

Figure 11.3. Non load-bearing panels, mounting.

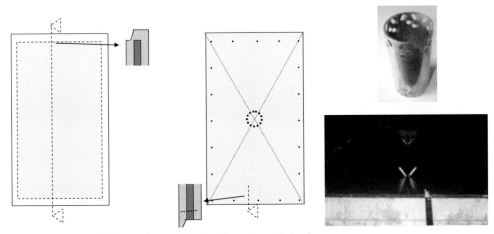

Figure 11.4. Sandwich panels: coupling inside and outside leaf.

During transport and mounting, the outside leaf may neither shift nor rotate compared to the inside one. To avoid that, both leafs were formerly coupled by a cast concrete perimeter. However, thermal bridging forced manufacturers to consider other solutions. Today, both leafs are coupled with a hollow stainless steel cylinder, if possible in the centre of gravity of the panel, a flat coupler close to the perimeter if the cylinder is not in that centre, and, ties all around (Figure 11.4).

11.2.2 Building physics: heat, air, moisture

11.2.2.1 Air tightness

Air-tightness is not a problem at panel level. If nevertheless air leakage is noted, infiltration and exfiltration across the joints is usually the reason.

11.2.2.2 Thermal transmittance

In case the insulation is continuous all over the panel clear wall transmittances ranging from 0.6 down to 0.1 W/(m$^2 \cdot$ K) demand the insulation thicknesses of Table 11.1, from 4–5 cm up to 22–37 cm. The thicker the insulation, the higher the load on the coupling elements between the inner and outer leaf and the larger panel width. That must be considered when designing a panelized building enclosure and creates secondary costs, which have a true impact on the investments.

For panels with a cast concrete perimeter, thermal bridging turns the clear wall value into a useless number. Also a central stainless steel hollow cylinder introduces some thermal bridging, especially when the cylinder gets filled with concrete and careless mounting of the insulation during manufacturing leaves joints between the boards. Figure 11.5 shows the impact of a cast concrete perimeter and a central stainless steel hollow cylinder on the isotherms, whereas Table 11.2 lists whole wall thermal transmittances for a storey-high panel. The lower the clear wall thermal transmittance, the more a cast concrete perimeter lowers insulation efficiency. Despite this, manufacturers continue promoting such panels using the clear wall value. A central stainless steel hollow cylinder solution with flat coupler and perimeter ties performs much better. In fact, the gap between clear and whole wall then looks marginal.

Table 11.1. Insulation thickness (8 cm thick outside and 14 cm thick inside).

U_o-value W/(m²·K)	Insulation thickness m			
	MW	EPS	XPS	PUR
0.6	0,05	0,05	0,04	0,03
0.4	0,08	0,08	0,07	0,05
0.2	0,17	0,18	0,15	0,11
0.1	0,35	0,37	0,31	0,22

Table 11.2. Whole wall thermal transmittance of a 3.6 m height and 2.4 m long panel.

Wall: 6 cm concrete/ mineral wool (d)/6 cm concrete	U_o W/(m²·K)	ψ W/(m·K)	U W/(m²·K)	$\Delta U/U_o$ %
With cast concrete perimeter frame (12 m)				
d = 4 cm	0.82	0.55	1.58	93
d = 8 cm	0.45	0.55	1.21	169
d = 12 cm	0.31	0.51	1.02	229
d = 16 cm	0.24	0.48	0.90	275
		χ W/K		
With insulation filled central stainless steel hollow cylinder (1st number in column 3) flat coupler and perimeter ties (2nd number in column 3)				
d = 4 cm	0.82	0.17/0.078	0.98	19.5
d = 8 cm	0.45	0.15/0.044	0.55	22
d = 12 cm	0.31	0.13/0.031	0.38	23
d = 16 cm	0.24	0.11/0.024	0.29	21

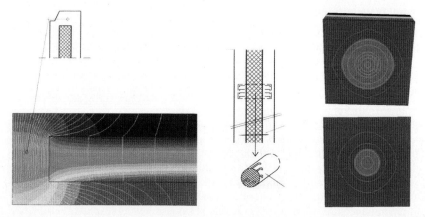

Figure 11.5. Manufactured sandwich panels, on the left the isotherms for a cast concrete perimeter frame, on the right the isotherms at the inside surface for a central stainless steel cylinder, down filled with insulation, up filled with concrete (in the three cases for 0 °C outdoors (blue) and 20 °C indoors (red)).

11.2.2.3 Transient response

Table 11.3 summarizes temperature damping, dynamic thermal resistance and admittance of a sandwich panel insulated with 6 and 19 cm EPS, assuming thermal bridge effects of the central stainless steel hollow cylinder, the flat coupler and the perimeter ties are marginal. The panels clearly perform excellently with a temperature damping exceeding 15, dynamic thermal resistance way above $1/U_o$ and the admittance clearly higher that half the inside surface film coefficient (7.8 W/(m²·K)). But, such panels are mostly used to construct office buildings, where the percentage of glass in the facades is high and inside partitions so light (hung acoustical ceilings, raised floor finishes, light-weight walls) that damping properties of the opaque facade parts do not matter anymore when considering overheating risk.

Table 11.3. Manufactured sandwich panel: temperature damping, dynamic thermal resistance, admittance (1-day period).

Panel	Temperature damping		Dynamic thermal resistance		Admittance	
	–	faze, h	m²·K/W	faze, h	W/(m²·K)	faze, h
6 cm EPS, $U = 0.57$ W/(m²·K)	40.9	9.2	6.2	7.9	6.6	1.3
19 cm EPS, $U = 0.20$ W/(m²·K)	120.8	10.1	18.3	6.6	6.6	1.3

11.2.2.4 Moisture tolerance

Wind drive rain

Precast concrete is hardly capillary, the capillary water sorption coefficient (A) being less than 0.018 kg/(m²·s$^{0.5}$). Impinging rain thus quickly runs off:

$$t_r = 0.000162/g_{ws}^2 \tag{11.1}$$

t_r being the moment run-off starts compared to the begin of the wind-driven rain event (in s) and g_{ws} wind driven rain intensity in kg/(m²·s). Fingering then leads to streaked soiling of the panels. Where there is run-off, dust is removed while elsewhere it accumulates. Suction of the fingering run-off by the concrete proceeds so slowly that moisture content hardly passes the critical one ($w_{cr} \approx 100$ kg/m³), see durability. Anyhow, joints between elements and between elements and windows may get heavily water-loaded, thus demanding excellent sealing.

Mould and surface condensation

With a clear wall thermal transmittance below 0.6 W/(m²·K), temperature ratio should stay above 0.7 and mould likeliness below 1/20 even behind cupboards against the outer wall. That also holds for surface condensation. However, things may go wrong with cast concrete perimeter panels, see 'thermal bridges'.

Interstitial condensation

Interstitial condensation is not an issue as more humid concrete – in the case being the outer leaf – has a much lower vapour resistance factor than air-dry concrete – here the inside leaf.

Because of that, a sandwich panel is an exemplary case of good assembly design: the inside leaf being the most, and the outside leaf being the least vapour tight. If a Glaser calculation shows deposit at the backside of the outside leaf, nothing more than a cyclic change in hygroscopic moisture content will be noted in reality.

11.2.2.5 Thermal bridging

Lowest temperature ratio inside along a cast concrete perimeter drops below the one measured centrally on double glazing. By that, mould risk increases drastically. For sandwich panels with a stainless steel hollow cylinder, flat coupler and perimeter ties, the lowest temperature ratio hardly deviates from what the clear wall thermal transmittance shows.

11.2.3 Building physics: acoustics

Sound transmission loss of manufactured concrete sandwich panels is truly sufficient for the windows to determine the facade's sound insulation.

11.2.4 Durability

As Table 11.4 shows, even for small insulation thicknesses the inside and outside leaf experience widely different thermal loads. If too stiffly coupled, important stresses can develop in the concrete. Assume a panel is manufactured at 10 °C. Once mounted, the outside leaf warms to a mean temperature of 48 °C during a hot day in the moderate North-Western European climate, the inside leaf to a mean temperature of only 23.4 °C. If perimeter coupling is infinitely stiff and shrinkage gives no embedded stress, tensile stress in the inside leaf will touch 3.4 MPa, i.e. approximately 1/10 of the compression strength of normal concrete. During a cold winter day the mean temperature of the inside leaf drops to 14.6 °C, and the outside leaf to −14.4 °C. Tension in the outside leaf then reaches 6.7 MPa, a value beyond the tensile strength of normal concrete. Spread cracking is thus not hypothetical. Cracks now act as preferential capillary paths. If they touch the bars, moisture content there will increase. This and easier CO_2 diffusion across empty cracks will accelerate carbonatation of the concrete with steel corrosion risk in the end.

Table 11.4. Manufactured sandwich panel: temperature load in the in- and outside leaf.

Panel, from out- to inside:	Leaf	Temperature variation, °C		
8 cm concrete EPS 14 cm concrete		Mean, cold winter/ hot summer day	Cold winter day	Hot summer day
EPS15, $d = 6$ cm	Inside	9.5	0.6	0.8
	Outside	61.5	19.7	27.7
EPS15, $d = 19$ cm	Inside	7.2	0.2	0.2
	Outside	62.5	20.0	29.1

Hygric loading is subordinate to the thermal loading. Concrete is quite hygroscopic. Because of that, the outside leaf will stay moderately moist in cool but humid climates while rain run-off is hardly sucked, at least as long as spread cracking is absent. For a concrete outside leaf capillary wetness over some 5 cm demands two days of continuous run-off. Even then, the difference in moisture content with the hygroscopic humid thickness left is so limited that elongation and bending remain hardly notable. Or, once shrinkage is completed, hygric loading becomes a phenomenon of secondary importance.

Cracking due to shrinkage and thermal loading is moderated by embedding a fine-meshed reinforcement in the inside and outside leaf and keeping the hardening concrete at high relative humidity during manufacturing.

11.2.5 Fire safety

Even with EPS as insulation, the fire resistance of manufactured concrete sandwich panels is excellent, on condition anyhow the EPS is fire retarding and the joints between panels fireproof. It is the windows that figure as ways for flame spread along panelized building facades. Therefore the requirement that the developed length of the opaque part between two successive floors must equal 1 meter or more is truly important.

11.2.6 Maintenance

Rain run-off defines the way a panelized building facade collects dirt. One sometimes tries to retard run-off by facade reliefs. Soiling, although more equal, may even be accelerated by that. Regularly cleaning is the only recourse. The high hygroscopic moisture content of concrete also increases sensitivity for algae growth. Treatment with an algae-killing product offers temporarily relief. For corroding reinforcement bars and spalling concrete there is only one solution: removing the concrete top layer, treating the bars with a corrosion inhibitor and protecting the whole with a suitable repair mortar.

11.3 Design and execution

As the analysis above shows, an affordable and durable panelized facade requires a modular design, allowing appropriate repetition to limit the number of different panels and to keep the lot per type large. Mineral wool, EPS, XPS and PUR are suitable as insulation materials. Inside and outside leaf demand a corrosion resistant coupling that assures strength and stiffness while minimizing thermal bridging. The best solution to date is the hollow stainless steel cylinder, possibly in the centre of gravity of the panel, a flat coupler close to the perimeter if the cylinder is not in that centre and glass fibre ties all around the perimeter.

Soiling and aging by uncontrollable cracking is curbed by using a well scaled concrete mixture with low water/cement factor, embedding welded mesh reinforcement in both leafs, assuring enough concrete cover and curing the concrete in an environment with high relative humidity. One of course also has to follow the concrete standards in all this.

11.4 References and literature

[11.1] TI-KVIV (1980). *Cursus Thermische isolatie* (in Dutch).

[11.2] Frick, Knöll, Neumann, Weinbrenner (1987). *Baukonstruktionslehre, Teil 1.* B. G. Teubner, Stuttgart (in German).

[11.3] Cziesielski, E. (1990). *Lehrbuch der Hochbaukonstruktionen.* B. G. Teubner, Stuttgart (in German).

[11.4] ASHRAE (2011). *Handbook on HVAC Applications, Chapter 44.* Atlanta.

[11.5] Hens, H. (2011). *Prefab wandelementen, Koudebrugwerking bij de sandwich elementen W200 en W201.* BPH_Consult bvba, Report 2011/17 (in Dutch).